W0227454

Plant Tissue Culture Manual
Supplement 1, February 1992

INSTRUCTIONS FOR SUPPLEMENT 1

Preliminary pages
Replace: Table of Contents

Section A
Add after Chapter A7:

Chapter A8: D. Valvekens, M. van Lijsebettens and M. van Montagu/*Arabidopsis* regeneration and transformation (root explant system)

Chapter A9: A.D. Krikorian and D.L. Smith/Somatic embryogenesis in carrot (*Daucus carota*)

Chapter A10: G. Spangenberg and H.-U. Koop/Low density cultures: microdroplets and single-cell nurse cultures

Section B
Add after Chapter B8:

Chapter B9: C.A. Rhodes and D.W. Gray/Transformation and regeneration of maize protoplasts

Section C
Add after Chapter C5:

Chapter C6: M.C. Coleman and W. Powell/Virus elimination and testing

Section D
Add after Chapter D6:

Chapter D7: P.K. Saxena and J. King/Isolation and uptake of plant nuclei

End matter
Replace: Index

Kluwer Academic Publishers

P.O. Box 17, 3300 AA Dordrecht, The Netherlands

Dear Reader

We would very much appreciate receiving your suggestions and criticisms on the *Plant Tissue Culture Manual*. They will be most helpful during our preparations for future supplements.

Would you please answer the questions listed below, and send your comments with any further suggestions you may have, to *Dr. M. Brewis* at the above-mentioned address.

Thank you for your assistance!

Dr M. Brewis
Publisher

— —

PLANT TISSUE CULTURE MANUAL

1. What errors have you found? (list page numbers and describe mistakes)
2. What protocols do you find to be confusing or lacking in detail? (list chapter numbers and page numbers and describe problems)
3. What protocols do you feel should be replaced in future supplements with newer (better) methods?
4. What new topics or other material would you like to see included in future supplements?

Please print or type your answers in the space below, and continue overleaf.

Name: Date:

Address:

PLANT TISSUE CULTURE MANUAL SUPPLEMENT 1

PLANT TISSUE CULTURE MANUAL

Supplement 1

Edited by:

K. LINDSEY

Plant Cell & Molecular Biology Group
Leicester Biocentre, University of Leicester, U.K.

Springer Science+Business Media, B.V. 1992

Library of Congress Cataloging-in-Publication Data

Plant tissue culture manual : fundamentals and applications / edited
  by K. Lindsey.
      p.   cm.
   Includes bibliographical references and index.
   ISBN 0-7923-1115-9 (acid-free paper)
   1. Plant tissue culture--Laboratory manuals.   I. Lindsey, K.
QK725.P587 1991
581'.0724--dc20   90-26765

ISBN 978-0-7923-1319-9

ISBN 978-0-7923-1319-9   ISBN 978-1-4899-3778-0 (eBook)
DOI 10.1007/978-1-4899-3778-0

Printed on acid-free paper

Contents

Preface

---

* Included in Supplement 1.

SECTION C: PROPAGATION & CONSERVATION OF GERMPLASM

---

* Included in Supplement 1.

SECTION D: DIRECT GENE TRANSFER & PROTOPLAST FUSION

---

* Included in Supplement 1.

Plant Tissue Culture Manual **A8**: 1–17, 1992.
© 1992 *Kluwer Academic Publishers.*

Arabidopsis regeneration and transformation (Root Explant System)

DIRK VALVEKENS, MIEKE VAN LIJSEBETTENS & MARC VAN MONTAGU
Laboratorium voor Genetica, Universiteit Gent, K.L. Ledeganckstraat 35, B-9000 Gent, Belgium

Introduction

The crucifer *Arabidopsis thaliana* has become widely used as a model system for plant molecular biology [1]. In this chapter we describe a simple and highly reproducible procedure for regeneration and transformation of *Arabidopsis thaliana* (L.) Heynh. [1, 2]. For this method the explant source is the root system of the plant.

After a short treatment with 2,4-dichlorophenoxyacetic acid (2,4-D) and subsequent incubation on a medium with a high concentration of N^6-(2-isopentenyl)adenine (2ip), *Arabidopsis* roots give rise to a vigorous bush of shoots over their entire length after only three weeks (Fig. 1). These R1 shoots can then be transferred to a plant growth regulator-free medium to produce seeds three weeks later. The aseptic R2 seeds can immediately be germinated on a plant medium without any pretreatment.

The most widespread approach for creating transgenic plants is the *Agrobacterium* system [3, 4]. Therefore, we incorporated our regeneration method with *Agrobacterium tumefaciens* infection to develop an efficient and rapid transformation procedure using kanamycin (Km) selection. Towards the end of the 2,4-D treatment of the root explants, *Agrobacterium tumefaciens* which contain T-DNAs that carry an antibiotic resistance marker gene can be cocultivated with the root explants. Transgenic T1 shoots can then be selected on the 2iP medium by the addition of an antibiotic that can be inactivated by the protein product encoded by the antibiotic-resistant marker gene together with an antibiotic which prevents *A. tumefaciens* from overgrowing the root explants (Fig. 2). Using this strategy, transgenic T2 seeds can routinely be obtained within three months of tissue culture.

We have also described a rapid nondestructive screening procedure that allows transformed and untransformed plants to be distinguished (Fig. 3). This method is extremely valuable for analyzing the segregation of T-DNAs in relation to other markers among the offspring of genetic crosses.

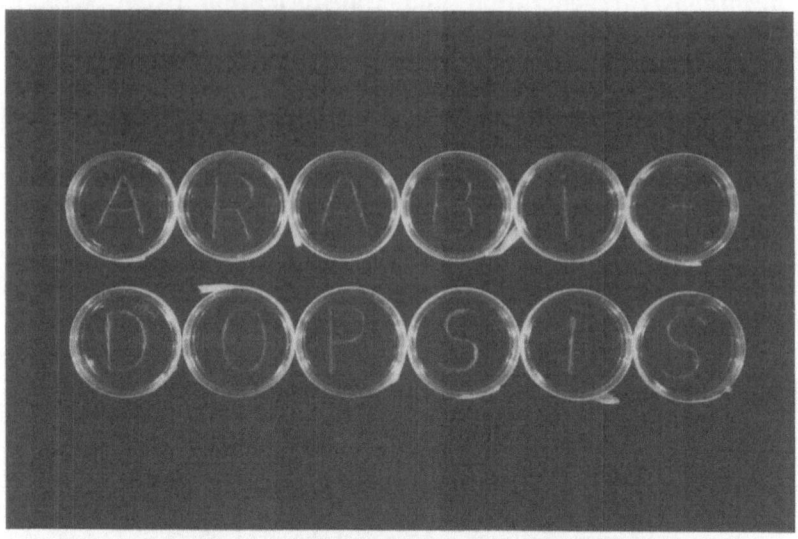

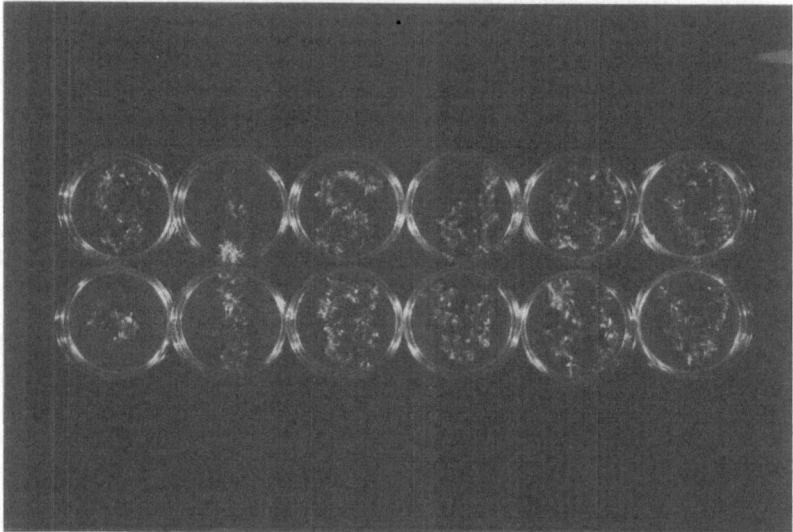

Fig. 1. 2,4-D-induced roots of *Arabidopsis thaliana* (a) give rise to a vigorous bush of shoots over their entire length after only three weeks of incubation on a medium with a high concentration of N^6-(2-isopentenyl)adenine (2ip) (b).

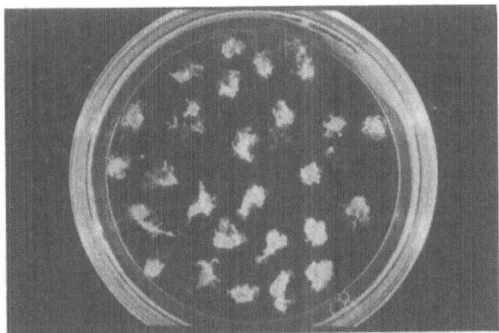

Fig. 2. After *Agrobacterium* infection transgenic T1 shoots are selected on a medium containing kanamycin and vancomycin.
Every C24 root explant gives rise to at least one transgenic shoot.

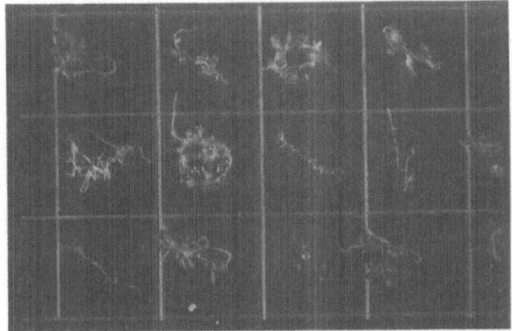

Fig. 3. Genetic test for marker gene expression in T2 progeny of transgenic lines.
Transgenic *Arabidopsis* T2 plants were identified by regenerating root segments on a kanamycin-containing medium. Eight out of the 12 root explants shown are resistant to kanamycin and produce numerous green shoots. The four remaining root explants originate from kanamycin-sensitive plants and do not produce any shoots.

Procedures

Seed sterilization and growth of seedlings
Steps in the procedure

1. Place seeds in 70% ethanol for two minutes (e.g. in a disposable, sterile 10 ml tube). Remove EtOH with a pipette.
2. Place seeds in 5% NaOCl/0.05% Tween 20 for 15 minutes. Shake regularly. Remove the sterilizing solution with a pipette.
3. Wash seeds in sterile, distilled water 5 times.
4. After the last wash, keep the seeds in 0.5—1.0 ml water. Take them up with a 2 ml pipette. Put drops with seeds on GM. Disperse the seeds homogeneously with the pipette tip.
5. Incubate the seeds on GM in a culture room for 7 days.
6. Transfer cotyledonary seedlings individually to fresh GM in 15 cm Petri dishes (Falcon® 1013) (15—30 seedlings/dish).

Notes
4. The seeds of some *Arabidopsis* ecotypes require a cold treatment prior to germination. In this case, the seeds can be stored dry for at least 7 days at 4 °C or can be incubated for 4 days at 4 °C on GM after sterilization.
5. The culture room should be operated with a 16 hour light/8 hour dark cycle and a temperature of 20 °C.
6. Sterilized seeds can also be immediately transferred individually to 15 cm Petri dishes; however, in our experience, germinated seedlings are easier to manipulate compared to seeds.

Regeneration from roots

For both regeneration and transformation *white* roots of aseptically grown *Arabidopsis* plants of any age can be used. After 8 weeks of growth, however, the roots will begin to turn green and should not be used.

Steps in the procedure

1. Pull plantlets *gently* out of the agar (GM) using forceps (the whole root system will come out easily).
2. Put the plantlets in a sterile Petri dish. Cut off root system from the rest of plantlet so that no green parts remain attached to the root.
3. Incubate the root system on 0.5/0.05 agar. Take care that the root is entirely in contact with the agar.
4. Incubate for 4 days in a growth room.
5. Transfer the roots to 0.15/5 agar. Numerous green shoots will completely cover the root system after 3 to 4 weeks.
6. Transfer individual shoots to GM in 15 cm Petri dishes (Falcon® 1013).
7. Incubate the growing shoots in the growth room for *in vitro* seed production.

Notes
1. Roots isolated from GM will quickly dessicate in a laminar-flow bench. Dessication can be reduced by quick handling and/or by keeping the roots submerged in some liquid plant medium (e.g. 0.5/0.05). For recovering *Arabidopsis* roots from agar several alternative methods can be used.
   – Remove the aerial parts of the plants with scalpel and forceps so that only the roots remain in the agar. Turn over the entire agar slab and recover the roots from the agar using forceps.
   – The use of 0.6% agarose instead of 0.8% agar causes the roots to grow on top rather than into the agar, leading to even more efficient recovery of root tissue [5].
7. To obtain seeds *in vitro* never put more than three shoots in each Petri dish. The lid of the Petri dish should always be absolutely free of condensation as high humidity inhibits anther dehiscence and hence prevents fertilization. The temperature should never exceed 22 °C. Rooted regenerants are obtained with 50% efficiency and can eventually also be transferred to soil to set seed. However, both rooted and unrooted regenerants will produce seeds *in vitro* with equal efficiencies.

Transformation of root explants

The root transformation procedure has been developed using the ecotype "collection number C24" [2]. For C24 an overall transformation efficiency of 80% (root explants giving rise to seed-producing T1 transformants) can be expected (Fig. 2). We have also successfully applied the method for transformation of Columbia and Landsberg *erecta* ecotypes. For Columbia, only 20% of the transformed root explants give rise to seed-producing regenerants. Transformation and complementation of an *Arabidopsis gl1* mutant, ecotype Columbia, has been reported [6], using slight modifications of our protocol. Landsberg *erecta* roots can be *regenerated* as efficiently as C24 roots. Also the transformation efficiency at the callus level (root explants producing green transformed calluses) is usually 100%. However, the regeneration of these transformed calluses is both retarded and less efficient in comparison with C24, suggesting that either the antibiotics or the agrobacteria are responsible for this negative effect. The overall transformation efficiency (i.e., root explants producing transformed T2 seeds) is usually about 20% over a 4-month period. The Bensheim and RLD ecotypes have also been successfully used for transformation [5, 7].

Using the root transformation procedure, "escapes" (isolates with 100% bleached seedlings in T2 populations germinated on GM K50) occur typically at a frequency of between 5% and 10%. When plant media are renewed more frequently than described below (e.g. weekly), it is advisable to use 60-70 mg/l kanamycin to prevent a higher "escape" frequency.

Although we exclusively used kanamycin selection in our transformation experiments, the successful use of hygromycin, chlorsulphuron, and methotrexate as selective agents for transformation of *Arabidopsis* root explants have recently been reported [8].

Steps in the procedure

1. Pull plantlets *gently* out of the agar (GM) using forceps (the whole root system will come out easily).
2. Put the plantlets in a sterile Petri dish. Cut off the roots from the rest of the plantlets so that no green parts remain attached to the roots.
3. Incubate the roots on 0.5/0.05 agar. Take care that the roots are entirely in contact with the agar.
4. Incubate in the culture room for 3 days.
5. Stack 3–5 2,4-D-induced roots in a sterile Petri dish. Cut the roots into 0.5 cm explants.
6. Transfer the root explants to a Petri dish containing 10–20 ml of liquid 0.5/0.05 medium.
7. Add 0.5–1.0 ml of an overnight grown *Agrobacterium* culture (28 °C; 200 rpm; Luria broth [9]). Shake gently for about 2 minutes.
8. Blot root explants briefly on a sterile filter paper to remove most of the liquid, and then place them on 0.5/0.05 agar.

9. Incubate up to 50 explants per 9 cm Petri dish in the growth room for 2 days to allow *Agrobacterium* infection. After 2 days of coculture the explants will be completely overgrown by agrobacteria.

10. Transfer the explants to 10–20 ml of liquid 0.5/0.05 medium. Shake vigorously to wash off agrobacteria. Blot root explants for a few seconds on sterile filter paper. Transfer explants to 0.15/5 V750 K50 agar, taking care that the root explants are in close contact with the agar.

11. Incubate for 2 weeks in a growth room (after approximately 2–3 weeks tiny, green calluses appear on the yellowish root explants).

12. Transfer the explants to 0.15/5 V500 K50 agar. Two weeks later the green calluses start to form shoots. These shoots are often vitreous (watery) at first. This "vitreous" effect is due to the presence of vancomycin. It is advisable to lower the vancomycin dose down gradually to 250 or even 0 mg/l after 4–5 weeks of tissue culture.

13. Transfer morphologically "normal" shoots to GM in 15 cm Petri dishes (Falcon® 1013) to allow further development. Among several other types of plant containers, this type of Petri dish was found to give the most efficient shoot regeneration and seed production. They can easily be kept free of condensation compared with other containers and are higher (2.5 cm) than standard 15 cm Petri dishes.

Notes

6. – A "root *explant*" is composed of multiple root *segments* and can be considered as a transverse cutting of about 0.5 cm through an entire root system and *not* as a cutting through only one branch of a root system. When performing infections or washings the explants sometimes fall apart, but then 5–10 root segments are taken together for further incubation. The root explants should not be longer than 0.5 cm. Histological examinations show that *Agrobacterium* often enters the vascular cylinder of root segments at the cut surface. Hence the more cut surfaces, the more entrance sites for the bacteria.

   – For convenience with *Agrobacterium* infections and washings of root explants, little autoclaved plastic baskets (3–4 cm high, approximately 6 cm in diameter) are used which have a 100 μm nylon mesh in the bottom (Fig. 4). The root explants are put in these little sieves to be infected or washed. In this way the explants can be removed from the mesh very efficiently rather than "fishing" them out of the liquid medium. This reduces the time needed for infections and washings by at least a factor of 2.

7. Only *Agrobacterium* cultures that were grown in nonselective LB medium should be used.

11. During vancomycin counter-selection bacterial overgrowth of the root explants can occur. This problem arises when a fluid layer is present on top of the agar. Therefore always allow the 0.15/5 vancomycin/kanamycin media to dry completely (30 minutes) in the laminar-flow cabinet before closing them.

13. *In vitro* seed set of regenerants is positively influenced by low humidity, so condensation of the lid of the Petri dish should by all means be prevented. Condensation arises when the bottom of a plant container has a higher temperature than the top. This situation often occurs in culture room racks by the heat produced by the illumination source. Further, one should be careful to culture shoots at low density (maximum 3 per 15 cm Falcon® 1013 Petri dish) and at low temperature (approximately 20 °C). Temperatures above 22 °C should be avoided.

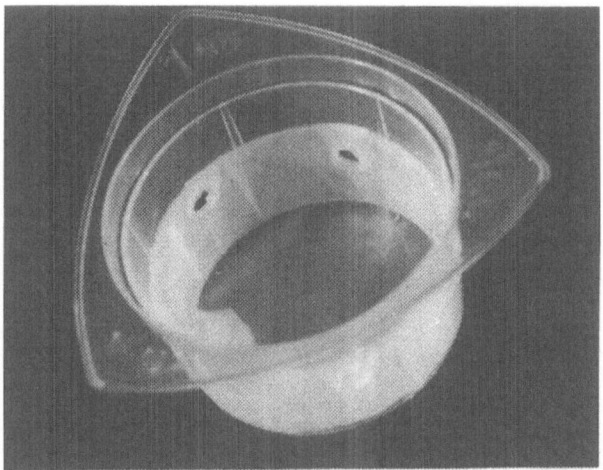

Fig. 4. Home-made basket containing a 100 µm mesh, which can be used for *Agrobacterium* infection and washing of *Arabidopsis* root explants.

Genetic tests for marker gene activity in T2 plants

The best criterion for demonstrating stable genetic transformation of a plant is the expression and Mendelian segregation of the selectable marker gene in the T2 progeny of T1 transformants. We describe both a selection and a screening procedure for transformed T2 seedlings.

Selection of transformed T2 plants
Steps in the procedure

1. Place T2 seeds on GM K50 agar in Petri dishes. Seal with Urgopore® tape.
2. Put the Petri dishes in the dark at 4 °C (refrigerator) for at least 4 days (stratification).
3. Incubate the germinating seedlings for 2 weeks in the growth room. Sensitive seedlings do not form roots nor leaves and develop white cotyledons; transformed seedlings are phenotypically normal.

Note
2. This step is not necessary if seeds were preserved for more than one week at 4 °C.

Screening for transformed T2 plants
Steps in the procedure

1. Remove young plants (4–6 leaves) from GM without damaging roots. Put plant in a sterile Petri dish.
2. Cut off half of the root system transversely.
3. Incubate the root explant on 0.5/0.05 agar. Put plantlet back onto GM. For this assay we use Falcon® 1013 (15 × 2.5 cm) Petri dishes. They contain a grid so that root explants and plantlets can be put in corresponding squares on either media.
4. After 4 days, transfer the root explants to 0.15/5 agar supplemented with 50 mg/l kanamycin in Falcon® 1013 Petri dishes (use corresponding squares).
5. About 4 days later, resistant roots can be distinguished from roots of sensitive plantlets with a high degree of accuracy. Roots of resistant plants bear many, long, straight root hairs whereas sensitive plants have at best a few shrunken ones. Another 5 days later, resistant roots turn green and begin to form shoots, whereas sensitive roots turn yellow and die. Greening of regenerating root tissue is a 100%-safe criterion for the determination of kanamycin resistance (Fig. 3).

Plant media

— GM (germination medium)
    1 × Murashige and Skoog salt mixture (Flow Laboratories)
    10 g/l sucrose
    100 mg/l inositol
    1.0 mg/l thiamine (stock 1.0 mg/ml)
    0.5 mg/l pyridoxine (stock 0.5 mg/ml)
    0.5 mg/l nicotinic acid (stock 0.5 mg/ml)
    0.5 g/l 2-(*N*-morpholino)ethanesulfonic acid (MES) (adjusted to pH 5.7 with 1 N KOH)
    8 g/l Difco Bacto agar
— 0.5/0.05 (callus-inducing medium)
    1 × Gamborg's B5 medium (without 2,4-D, kinetin, and sucrose) (Flow Laboratories)
    20 g/l glucose
    0.5 g/l MES (pH 5.7 adjusted with 1 N KOH)
    8 g/l Difco Bacto agar (for solid 0.5/0.05 medium)
    0.5 mg/l 2,4-D (stock 10 mg/ml)
    0.05 mg/l kinetin (stock 5 mg/ml)
— 0.15/5 (shoot-inducing medium)
    1 × Gamborg's B5 medium (without 2,4-D, kinetin and sucrose) (Flow Laboratories)
    20 g/l glucose
    0.5 g/l MES (pH 5.7)
    8 g/l Difco Bacto agar

5 mg/l 2ip (stock 20 mg/ml)

0.15 mg/l indole-3-acetic acid (IAA) (stock 1.5 mg/ml)

— 0.15/5 V750 K50. As 0.15/5, but supplemented with 750 mg/l vancomycin (Vancocin HCl; Eli Lilly) (stock 200 mg/ml) and 50 mg/l kanamycin sulphate (Sigma) (stock 50 mg/ml)

— 0.15/5 V500 K50. As 0.15/5, but supplemented with 500 mg/l vancomycin and 50 mg/l kanamycin

— GM K50. As GM, but supplemented with 50 mg/l kanamycin

Notes

— All media should be autoclaved for 15 minutes at 121 °C. Vitamins are added and the pH adjusted to 5.7 before autoclaving. Plant growth regulators and antibiotics are added after autoclaving and cooling of the media to 55 °C.

— Plant growth regulators are dissolved in dimethylsulfoxide (DMSO) and added to the autoclaved media without any further treatment. Antibiotics are dissolved in water and subsequently filter-sterilized using a 0.22 µm Millipore filter.

— The choice of the agar is extremely important for efficient regeneration. Several agars that we have tested proved to contain inhibiting agents or simply to have a bad structure, resulting both in slower regeneration and a severely reduced regeneration efficiency.

— Falcon® Optilux 1005 (10 cm × 2.0 cm) Petri dishes are used for pouring 0.5/0.05, 0.15/5, 0.15/5 V750 K50, and 0.15/5 V500 K50 agars. Falcon® 1013 (15 cm × 2.5 cm) Petri dishes are used for pouring the GM and GM K50.

— It is very important to allow the media to solidify completely in a laminar flow and preferentially to wait 15 minutes further before closing the Petri dishes in order to prevent the formation of a liquid layer on top of the medium.

— All Petri dishes were sealed with Urgopore® (Chenove, France) medical gas-permeable tape. Many other types of tape were found to have a negative effect on the regeneration efficiency.

References

1. Meyerowitz, E.M. (1989). *Arabidopsis*, a useful weed. *Cell* 56, 263–269.
2. Valvekens, D., Van Montagu, M., and Van Lijsebettens, M. (1988). *Agrobacterium tumefaciens* -mediated transformation of *Arabidopsis* root explants using kanamycin selection. *Proc. Natl. Acad. Sci. USA* 85, 5536–5540.
3. Gheysen, G., Herman, L., Breyne, P., Van Montagu, M., and Depicker, A. (1989). *Agrobacterium tumefaciens* as a tool for the genetic transformation of plants. In *Genetic transformation and expression*, L.O. Butler, C. Harwood, and B.E.B. Moseley (Eds.). Andover, Intercept, 151–174.
4. Hooykaas, P.J.J. (1989). Transformation of plant cells via *Agrobacterium*. *Plant Mol. Biol.* 13, 327–336.
5. Chaudhury, A.M., and Signer, E.R. (1989). Non-destructive transformation of *Arabidopsis*. *Plant Mol. Biol. Reporter* 7, 258–265.
6. Herman, P.L., and Marks, M.D. (1989). Trichome development in *Arabidopsis thaliana*. II. Isolation and complementation of the *GLABROUS1* gene. *Plant Cell* 1, 1051–1055.
7. Márton, L. and Browse, J. (1991). Facile transformation of *Arabidopsis*. *Plant Cell Reports* 10, 235–239.
8. Honma, M.A., Waddell, C.S., and Baker, B. (1990). Development of an *Ac/Ds* transposon tagging system in *Arabidopsis*. Abstract presented at the Fourth International Conference on *Arabidopsis* Research, Vienna (Österreich), p. 12.
9. Miller, J.H. (1972). *Experiments in Molecular genetics*. New York, Cold Spring Harbor Laboratory.

Plant Tissue Culture Manual **A9**: 1–32, 1992.
© 1992 *Kluwer Academic Publishers.*

Somatic Embryogenesis in Carrot *(Daucus carota)*

ABRAHAM D. KRIKORIAN & DAVID L. SMITH
*Department of Biochemistry and Cell Biology, State University of New York at Stony Brook,
Stony Brook, New York 11794–5215, USA*

Introduction

The culture of carrot cells in liquid suspension dates from 1953 and the recognition of their totipotency from 1956 [10]. By 1962 it was feasible to maintain in the laboratory, routinely, liquid cultures, heterogeneous as to their unit size, but in which large numbers of embryos readily developed from suspended cell clusters and single cells [8]. By this time, the role of synergistic combinations of the growth-promoting complex as it occurs in coconut water with auxins such as naphthaleneacetic acid (NAA) and 2,4-dichloro-phenoxyacetic acid (2,4-D) had become well-known [11, 23]. Moreover, the advantages to be gained in some otherwise morphogenetically recalcitrant cell cultures, of sequential treatments with different growth-promoting complexes and systems, became appreciated [23]. By these general means it was shown that a number of umbelliferous plants (family Apiaceae) and species or cultivars from other families, could yield cells and somatic embryos which in turn could give rise to whole plants. But when the main sequence of embryogenic development of carrot cells became known [24], it was found that its outcome could be greatly altered by the environmental conditions and the identity and mode of application of the growth and morphogenetic stimuli [11, 23].

As it became clearer that the innate totipotency of somatic (normally diploid) cells is a presumptively general property, the following problems emerged:

(a) cell cultures of some plants (natural species or cultivars) may prove to be so recalcitrant that somatic embryos either do not form or do so too infrequently for this to be of practical use;

(b) unlike the normal course of zygotic embryogenic development, which is very closely controlled in the environment of the ovule and the embryo sac, that of somatic embryogenesis is much more responsive to external conditions [23, 24]. In the outcome, therefore, somatic embryos may differ in varying degrees from their zygotic counterparts and are much more variable, even within a given *in vitro* culture, in both size and form. Therefore, it is a real problem for an investigator to adjust, and readjust, the environmental conditions and the composition of the ambient medium so that embryogenic cell cultures may resemble more closely the development of zygotes and also develop along their assigned course as predictably.

(c) since the various methods for isolating plant protoplasts are now well established [5], it should be possible to begin the somatic embryogenic process with naked somatic protoplasts, prepared from embryogenic carrot cells in culture. These should, in theory at least, resemble zygotes. However, to the present, it has proven unexpectedly difficult to promote, reliably, high fidelity somatic embryogenesis in free protoplast cultures of carrot (and any other species for that matter) even when prepared from highly embryogenic units [13, 24], that as such, will develop in high yield to well-formed somatic embryos and plantlets.

Totipotency in cultured cells

Somatic embryogenesis in cultured plant cells relates to the origin in small proembryonic clusters, cells or initials that give rise to shoot and root growing points. After this has occurred, embryogenic and subsequent development is subject to all the environmental factors and interactions that beset whole plants.

Some definitions

In the early period, somatic embryos were frequently referred to as "embryoids". This was to distinguish them from zygotic embryos. Use of the word "embryoid" has diminished over the years and its complete abandonment would constitute no great loss. We refer to embryogenically competent cells or cell clusters as preglobular stage proembryos (PGSPs). This follows Halperin who first used the phrase in connection with somatic embryogenesis [6]. The term is fully justified since, if PGSPs are generated in the manner to be detailed in this section, they are, indeed, precursors to globular stage embryos. Others have preferred to call them embryogenic cell clusters, proembryonic masses (PEMs), or proembryonic globules and argue that the term PGSP begs the issue whether a single cell in the cluster gives rise to embryos, or whether a group of cells yields the next (globular) stage. The fact is that until the culture methods become standardized and better understood, these issues will, for us, remain moot points [3, 4, 8, 24]. Whatever term is used to designate these units (be they single cells or cell clusters of varying size), the terms are justified only if the next stage can be demonstrated as occurring. If something is a called a PGSP, it should only be called a PGSP if it can progress to a globular stage embryo. Similarly, the term globular stage embryo is preferable to the earlier and more common designation, proembryo. If it progresses to the heart stage, then one is justified in calling it a globular stage somatic embryo. Similarly, a heart-shape stage progresses to a torpedo stage and that, in turn, to the cotyledonary stage. Aberrant or abnormal embryos may be defined precisely in morphological terms if that is possible. In this laboratory, the term "neomorph" has been used to designate abnormal embryonal forms. This word was invented by Waris [26] to describe severely deformed seedlings that had been achieved *in vitro* by the use of anti-metabolites and amino acids. For us, neomorphs represent terminal products of inaccurate delivery or processing of various signals in the embryogenic or developmental process. A key feature of neomorph status, as it were, is that gene mutation is not involved. The term "terminal product" seems justified since the aberrant form (neomorph) does not progress directly to a plantlet. They invariably do not survive to become plantlets because their growing points are absent or ill-formed. On the other hand, explanted tissues from neomorphs can be induced to de-differentiate and, provided the environment is such that embryogenic signals are correct, cells from the neomorph can yield somatic embryos. These, in turn, can make plantlets. All this suggests that neomorphosis can be used as a tool to dissect the epigenetic and physiological aspects of developmental pathways.

Somatic embryogenesis protocols for carrot

The classic literature [1, 2, 6, 8, 10] reveals that the culture procedures were developed in a largely empirical fashion and leave something to be desired in terms of reliability. The fact remains, moreover, that despite the use of carrot to study somatic embryogenesis for a third of a century [10], and its being viewed by many as a routine and well-established procedure (so much so that it is frequently cited as the "model system"), there are fewer principles that can be grasped in terms of why a given procedure "works" than one might hope for. Because it was the first case where somatic embryogenesis was encountered *in vitro*, an extensive literature indeed exists [1, 2, 3, 4, 7].

Nevertheless, it will be helpful to recapitulate in some detail how embryogenic carrot cultures may be established using these nominally long-available procedures. But, since many laboratories have used various protocols, it is hoped that investigators will see opportunities to probe further both the "older" and "newer" methods to be presented and thus through comparative studies be able to gain better understanding of the full range of controls that must be in place in developing and growing plants.

Initiation of primary cultures

Source of explants

Cells derived from carrot root secondary phloem explants are embryogenically competent [23, 24]. Similarly, zygotic embryos from developing or mature carrot "seeds" (botanically they are "mericarps" or fruits) also yield vigorous cultures capable of organizing. Nevertheless, a real problem in all of this is to have materials which are not only embryogenic but which maintain over time a high level of response to the stimuli which allow them to organize. It may come as a surprise to some investigators seeking to work with carrot that considerable work must be done to enable one to select clones that respond in a predictable and desirable way. In this laboratory, over the years, selection procedures have been applied to cultures (primarily suspensions) initiated from the secondary phloem and even the cambium of cultivated carrot root [24]. They have also been applied to cultures started from explanted tissues of aseptically germinated carrot seedlings such as hypocotyls. Whereas only minor differences in morphogenetic response have been detected between cultures derived from cambium or secondary phloem, differences are detectable in cultures derived from seedlings of several cultivars. For instance, although Gold Pak, Royal Chantenay, Imperator Long, Scarlet Nantes, Spartan Delight, Hi Pak etc. all respond and produce somatic embryos, they have different requirements [24] and whether one seeks to study one or the other cultivar will depend on one's objectives. This is an important point and underscores the view that "model" systems are dependent on objectives.

"Seeds" or mericarps

Two types of carrot "seed" have generally been used for establishing embryogenic or totipotent carrot cultures: (1) *Daucus carota* var. *carota* L. (the Wild Carrot or Queen-Anne's-Lace) and *Daucus carota* var. *sativus* (the cultivated carrot) [17]. In the former case, the plants grow wild in open fields, meadows and roadsides, and "seeds" may be readily harvested. It is essential to establish on the spot, however, that the mericarps contain embryos. A common pest of carrot is the "tarnished plant bug" (*Lygus* spp.). This insect is able to selectively suck out the developing embryo in a "seed" without leaving signs of obvious damage. A mericarp presumed to be healthy may contain no embryo [14]!

"Seed" sterilization

The two following procedures provide, step-wise, instructions for the surface decontamination or disinfestation of carrot "seeds"- assuming that no microorganisms are present within the boundaries of the fruit wall.

Procedure 1 is the easiest and quickest method if the "seeds" are to be plated out for germination and if filtering pans with autoclavable (stainless steel) sieves are available. (In the USA, Cellector Tissue Sieves by E-C Apparatus 3831 Tyrone Blvd, North St Petersburg, FL 33709 have been satisfactory. See Fig. 1 on p. 17 for the appearance of these.)

Procedure 2 is the preferred method if zygotic embryos are to be dissected from the "seeds". Each procedure will disinfest approximately 500 "seeds" (there some 1000 per 10 ml glass beaker).

Procedure 1

Steps in the procedure

1. Sterilize one 250 ml Erlenmeyer flask containing 50 ml of glass distilled water (or equivalent), two 500 ml Erlenmeyer flasks containing 250 ml each of water, filtering pans with sieves comprising either a # 60, # 50, or # 40 mesh screen (see Table 1, p. 12) supported on a 400 ml glass beaker (protect filtering pan and beaker with aluminum foil), and one 100 mm diameter glass Petri dish per batch of 100 "seeds" being disinfested.

2. Add 50 ml of commercial bleach (e.g. Clorox which contains about 5.25 percent w/v sodium hypochlorite) and a drop of a surfactant such as Tween 20 into the 250 ml Erlenmeyer flask containing 50 ml of water.

3. Carefully pour the desired number of "seeds" into the 250 ml flask directly from the seed package, or from a small (10 ml) beaker and cover. Five ml is about 500 "seeds".

4. Briskly swirl every 15 minutes for 1.5 hours.

5. Pour "seeds" onto the filtering pan sieve and allow to drain.

6. Hold filtering pan over a one liter beaker and slowly pour the contents of both 500 ml flasks over the "seeds" to rinse off the hypochlorite.

7. Turn the filtering pan over and gently tap out the "seeds" into the bottom of the glass Petri dishes, or, alternatively use a sterile scoop to transfer the "seeds" into the Petri dishes.

Procedure 2

Steps in the procedure

1. Sterilize seven 250 ml Erlenmeyer flasks, one containing 50 ml, the others 100 ml each of distilled water and one glass Petri dish (100 mm diameter) per batch of 100 "seeds".
2. Add 50 ml of commercial bleach and 1 drop of a surfactant to the 250 ml flask containing 50 ml of water.
3. Carefully pour the desired number of "seeds" into the flask directly from the seed package, or from a small beaker and cover.
4. Briskly swirl every 15 minutes for 1.5 hours.
5. Carefully pour out the solution after flaming the neck of the flask. Some "seeds" will fall out but not many if you're careful. Again flame the neck of the flask.
6. Pour in 100 ml of water from the other flasks one at a time repeating step 5 after each addition. The last addition of water remains in the flask to allow the "seeds" to soak.
7. Soak the "seeds" for at least two hours, or as long as overnight. Approximately six hours is optimum.
8. Drain off the water.
9. Using a sterile scoop, place approximately 50 "seeds" into a sterile Petri dish for dissection under a stereoscopic microscope, for plating or germination.
10. Embryos may be excised from the "seeds" with two pairs of fine point forceps (e.g. Dumont # 3) or a single pair of forceps and a scalpel. This can be done by nicking the pointed end of the "seed" and pushing the embryo out from the other end. With practice, this procedure can be carried out readily. The embryos are then placed on an appropriate culture medium. For culturing either "seeds", embryos or seedlings, any of a number of culture vessels may be used.

Initiation of primary cultures

Media used in initiation of seedlings as a source of primary explants

The media most commonly used for germination of seeds and culture of embryos are one-half strength basal Murashige and Skoog salts (abbreviated in this laboratory as $B_{MS1/2}$), "vitamins" (micro-organics) [12] and iron chelate [15], or one-half strength modified White's salts (abbreviated as $B_{W1/2}$), iron, and "vitamins" [16], each supplemented with 0.5% w/v sucrose. B_{MS} is usually adjusted to pH 5.4; B_W to pH 6.2. B_W can profit from addition of 200 mg/l of casein hydrolysate (enzymatic digest or acid hydrolyzed). Normally B_{MS} is supplemented with 3% w/v sucrose; B_W with 2%. The low amount of sucrose (0.5%) used in the germination of seedlings has been shown to give a more uniform response. Indeed, this is added to permit disclosure of any microbial contaminants. Coconut water, abbreviated CW, (10% by volume) has frequently been used as an additive to B_W. It is also sometimes used in conjunction with full strength Murashige and Skoog salts, iron and vitamins (B_{MS}), especially with excised embryos. Coconut water seems to produce a more vigorous seedling.

Media utilized in initiation of callus

If callus is desired from the "seed", embryo or seedling, growth regulators such as NAA or 2,4-D (either at 2 mg/l, i.e. 10.8 μM or 9.0 μM respectively) may be added to the full strength salts. Seedlings are allowed to germinate in darkness at about 22 °C on basal medium. After about 10 days, when seedlings have reached a height of about 8 to 10 cm, including the root, hypocotyl and two (or three) etiolated cotyledons, they are suitable for culture. (That is not to imply that earlier stages or light-grown ones cannot be used). The seedlings are usually cut into sections approximately one cm long, and are then gently wounded with a scalpel along their length, and then placed in or on the culture medium. The sections may be placed on semi-solid medium in culture vessels or Petri dishes or directly into liquid medium.

Normally, an inoculum should comprise the tissue from a single seedling. This implies that a clonal population is being sought. The disadvantage of a culture derived from a single source is that it takes a longer period to obtain a large enough population of cells to work with but it provides the only means whereby one can establish specifics of performance of a given clone.

Since carrot suffers from inbreeding depression, there is little opportunity at present to work with isogenic stocks. By using clonal lines, one can, hopefully, come to a better understanding of control processes in embryo development. The vast majority of embryogenic cultures of carrot (and other species as well) have not, however, been clonally derived and hence represent mixtures of cells that may have very different characteristics.

Initiation of embryogenic cultures is best carried out in darkness, although dim light will be satisfactory in some cases. This will depend on the clone in question [24]. Within a week, or two at the most, signs of globular stage embryo production should be apparent. These will be discernible by their "pearly" white appearance. Their numbers and place of production on the primary explant will vary depending on how a culture is initiated. If a petiole or hypocotyl explant is used, for example, the cut or wounded ends will show the somatic embryogenic cells first.

Recognition of these small, pearly white and globular structures (globular stage embryos) is crucial to achieving an embryogenically active culture. These globular stage somatic embryos must be removed and multiplied via subculture under appropriate conditions to get a dense or vigorous culture, be it in suspension or on semi-solid medium. In addition to globular stage embryos giving rise to other globular embryos (proembryonic globules), they may shed, especially in liquid, embryogenic cell clusters (so-called proembryogenic masses or preglobular stage proembryos, PGSPs).

It is right at the beginning that choices in medium must be made, i.e. the exact auxin concentration to be used, whether or not to include coconut water or a chemically identified cytokinin in its stead, or whether to subculture in the same medium or to switch to another one, etc. Empirical evidence shows: somewhere in the culture protocol the use of basal medium plus an auxin is beneficial in yielding cells which can undergo embryogenesis more readily (apparent when the cultures are later evaluated for their embryogenic potential; this turns out to be "sooner rather than later"). Probably, the use of auxin at the outset is best, whether on semi-solid or in liquid medium. But if auxin is used at the outset, the medium must eventually be changed to include coconut water or a cytokinin, because with auxin by itself, the suspensions do not grow well for long, and indeed, are rarely maintainable with the right qualitative characteristics for more than three or four subcultures. That is, the suspension grown in medium supplemented only with auxin predominantly yields relatively large globular embryos, and there are few, if any, small isodiametric cells or tight clusters comprised of only a few (4–6) cells. These latter cell types are the ones that are most embryogenically responsive [24, 25].

Actively embryogenic suspensions have also been obtained over the years in this laboratory by placing primary explants of seedling sections directly into liquid medium with an auxin and coconut water. This can save at least a couple of weeks in establishing suspensions.

The basal salts medium of Murashige and Skoog [12] (i.e. a high nitrogen medium) promotes rapid proliferation of cells and is well suited to the initiation of growth from explants; it is also useful later as a medium in which totipotent cells may be placed to test their embryogenic potential. The B_{MS} may be inappropriate, however, for some cell lines and in these cases it is not possible to maintain the required level of competent cells that can develop into somatic embryos when placed in an auxin-free medium. Rather, the cells tend to grow into amorphous callus masses with no externally visible organization. Modified White's medium (B_W) [16] seems to be better suited for most routine and sustainable subcultures of embryogenic suspensions provided the appropriate growth regulators (e.g. CW, 2,4-D or NAA and casein hydrolysate) are added.

Procedures used for subculturing cells in suspension and evaluating them for their embryogenic potential

After the primary explants placed on semi-solid medium have callused and localized areas have produced globular stage embryos, the pearly white sectors may be transferred or subcultured onto semi-solid medium, or this embryogenic callus may be placed in liquid where, in time, it will produce a suspension. Seedling segments directly placed in small flasks usually produce a suspension in four to five weeks when the medium contains a cytokinin and auxin, and in approximately six weeks when it does not. Suspensions can be achieved earlier by combining contents of several flasks or increasing the number of units comprising an inoculum in a culture vessel but, as suggested above, it is better to obtain clonal lines for studies of development.

Routine subculturing of material grown in flasks may be done in the following way

Ideally, suspensions should be separated into fractions of known size by sieving (filtration) of the suspensions through a series of graded stainless steel or nylon sieves. Embryogenically competent suspensions are readily grown and multiplied as small cell clusters in various media, and, as stated above, these media may comprise any of several formulations comprising inorganic salts, sucrose, "vitamins", coconut water and an auxin, usually either 2,4-D or NAA. Lowering or removing the auxin from a totipotent cell culture-maintenance medium such as that of White and transferral to a hormone-free (H-F) one that utilizes the salts and vitamins of Murashige and Skoog [12] leads to the production of later stage somatic embryos [1, 2, 6].If coconut water is used, it and the auxins are usually removed, although the removal of auxin is by far the more important step. Presence of auxin inhibits progression of globular stage embryos to later stages of development.

Somatic embryo formation depends on a number of things including the line of carrot employed. In certain lines there is only a very low level of response, with unorganized clumps also being present in the suspension. This also tends to be the case when the auxin concentration is only lowered but not removed completely. Frequently, in a suspension that has had growth regulators removed via washing in H-F medium or subculture to a H-F medium, the newly formed embryos grow very rapidly and pass quickly through the embryo stages into plantlets. Another noteworthy fact is that with the removal of hormones, the cell culture tends to become very sparse and usually is lost after two passages on medium with no hormones. (More information about the use of H-F media is described later). This is probably due to the fact that somatic embryos do not (under those circumstances) retain their ability to slough off the new cells that would enable the maintenance of the suspension even as somatic embryos were being produced. Carrot suspensions in the presence of hormones normally comprise embryogenic cell clusters and the extent to which these can be recognized by a skilled investigator as embryogenic cells, and that may be equated with certain early stages of zygotic embryogenesis such as the globular stage embryo stage or the preglobular proembryo stage, is a moot point.

Be that as it may, development of cells and clusters into somatic embryos and plantlets can proceed readily with cells obtained by either of two procedures. 1) The fraction that has passed through a crude filter such as that provided by a single or double layer of cheesecloth stretched over a 400 ml beaker and supported by a taught string may be used, or 2) further procedures may be adopted using the cheesecloth filtrate first, to yield a more refined fraction. As an example of the latter: the cheesecloth filtrate can be poured through sieves of different mesh sizes such as # 100, # 200, # 400 (see Table 1).

Table 1. Some examples of screen sizes as used in the protocols presented. Note: The nominal dimensions of mesh (wires/inch) and pore size do not correspond exactly to US Bureau of Standards Sieve sizes but values are close.

Mesh (#)	Pore Size (μm)	US Standard Sieve (μm)
20	860	841
30	520	545
40	380	420
50	280	297
60	230	250
80	190	177
100	140	149
200	73.7	74
300	45.7	not available
400	38.1	37
500	25.4	25

The resulting material may be collected in two different ways: (1) by allowing the filtrate to stand so that the cells settle, followed by pouring off the supernatant; or (2) after pouring material through a series of filtering pans, the material can be collected directly off the smaller mesh by transferring the residue on the mesh into a beaker with a "rinse" medium (comprising the medium into which transfer or platings/plantings are to be made), thus producing a suspension of cells and/or clusters in a particular size range, e.g. 74–140 μm. In either case, the material finally collected is poured into a capped centrifuge tube and centrifuged at low speed, approximately 100–300 rpm (2–$15 \times g$) for about 10 minutes. The supernatant is poured off and the remaining cells and clusters may be further washed in the centrifuge tube with the rinse medium, re-centrifuged, etc., or it may be put into a culture medium directly without carrying out additional washes. The suspension in the centri-fuge tube is brought up to volume using the same medium that is to be used as the test medium. The final volume to be achieved can be estimated from experience by judging (by use of graduated test tubes) the density of the cells. This may also be quantified by counting an aliquot of cells.

All this means that when a culture is active and hormone is provided in the maintenance phase, few if any late stage somatic embryos are produced. Accordingly, it is a relatively simple matter to maintain early stage embryonic suspensions via subculture. Hormone-maintained suspensions are poured through cheesecloth or sieves to achieve filtration. Inocula can be used from those fractions or the fraction which does best in a particular setting. A volume consisting of some 1 to 10 ml (some 1 g fresh weight at the higher volume) may be used to inoculate a flask which accommodates 225 ml of liquid medium. Since the auxins generally prevent further development or progression of the somatic embryogenic process, it follows that if 2,4-D or NAA is subsequently removed, numerous somatic embryos will form. Also, it is at the time of transferral to H-F medium that the largest number of somatic embryos are produced. These are, however, inevitably of different sizes and hence must be separated by sieving if they are to be suitable for detailed investigation.

Isolation and fractionation of embryos

Embryogenically competent suspensions of carrot containing mixed populations of materials, whether they be competent cells or cell clusters, or somatic embryos ranging from freely suspended preglobular stage proembryos or globular stage embryos to fully formed somatic embryos can be separated as follows. Fig. 1 (p. 17) provides an abbreviated schematic version of some of the key steps in the process.

Steps in the procedure

1. Flame the neck of the culture flask and allow to cool for about 30 seconds.
2. While the neck of the flask cools, remove an aluminum foil protective cover from a 250 ml glass beaker and place a # 20 (864 μm pore size) filtering pan over the beaker.
3. Pour the contents of the flask through the # 20 screen into the beaker. The largest somatic embryos and plantlets are retained on this size sieve.
4. Wash the sieve with a rinse medium consisting of the salts and vitamins and sucrose to flush through any smaller embryos remaining in the larger mass. Usually these larger embryos retained on the # 20 screen are discarded. A small amount of the embryo suspension from the original culture flask may be put aside in a beaker instead of passing the entire contents of the flask through the # 20 sieve. This material may be used later as a source of inoculum for a new maintenance culture flask.
5. After a few minutes, most of the embryos in the filtrate settle to the bottom of the beaker. Some of the upper filtrate can be decanted so that the volume of the filtrate does not become too large. This might occur because rinse medium is added following each sieving. This decanting step can be done at the same step throughout the procedure.
6. Repeat the same procedure of sieving by pouring the filtrate from the # 20 sieve through a # 40 sieve (380 μm) which also rests on a 250 ml beaker. The material retained on the # 40 sieve is washed with rinse medium.
7. Collect the filtrate in the beaker, cover with foil and set aside while the # 40 sieve with retained embryos is flipped over and placed on a sterile 400 ml glass beaker.
8. Pour rinse medium through this sieve to rinse off these embryos into the 400 ml beaker.
9. Pour the embryos then into a 35 ml conical centrifuge tube and label # 20–40. A small aliquot of this embryo suspension can be put into a small Petri dish (e.g. a Falcon plastic 1006, 50 mm diameter) for observation under an inverted microscope.
10. Pass the filtrate in the beaker from the # 40 sieve then through a # 60 sieve in a similar manner and rinse with fresh medium.
11. This filtrate is again set aside and the sieve with retained embryos again flipped over and the embryos collected in a 400 ml beaker, then poured into a centrifuge tube and labelled # 40–60.

12. Repeat this procedure with a #80 (190 µm) sieve and then a #100 (140 µm) sieve with embryos collected from each sieve, yielding #60–80 and #80–100 fraction.

13. Pour the filtrate that passes through the smallest sieve, the #100 sieve, into several centrifuge tubes and label < #100.

By this process embryos have now been collected in the centrifuge tubes in the following ranges: #20–40, 380–860 µm; #40–60, 380–230 µm; #60–80, 230–190 µm; #80–100, 190–140 µm; < #100, <140 µm.

Sieves of intermediate size can also be used, i.e. #30, 50, 70, etc. to achieve whatever size ranges are desired. Fewer sieves, of course, can be used to achieve larger size ranges.

The < #100 fraction is comprised of small embryos at the heart stage, as well as smaller units and cells. Thus, this fraction usually requires centrifugation to concentrate the embryos and cells. This is done by centrifugation at about 300 rpm for 10 minutes. The fraction of the larger sizes generally settles quickly and completely, and thus does not require centrifugation. The supernatant is then carefully poured off all the tubes and the embryos are re-suspended in fresh medium without hormones or coconut water. The amount of medium to be added is estimated by determining the density of the embryos. The embryo concentration at this point can be quantified by withdrawing an aliquot, say one ml, pipetting it (make sure it has an appropriately wide mouth) into a small Petri dish and counting the number of embryonic units. Care must be taken to ensure that a well-mixed sample is withdrawn, particularly in the case of the larger embryo sizes, since they rapidly sink to the bottom of the tube or pipette. When the count is made, the density can be adjusted further by adding more medium or units and a second count carried out, if necessary.

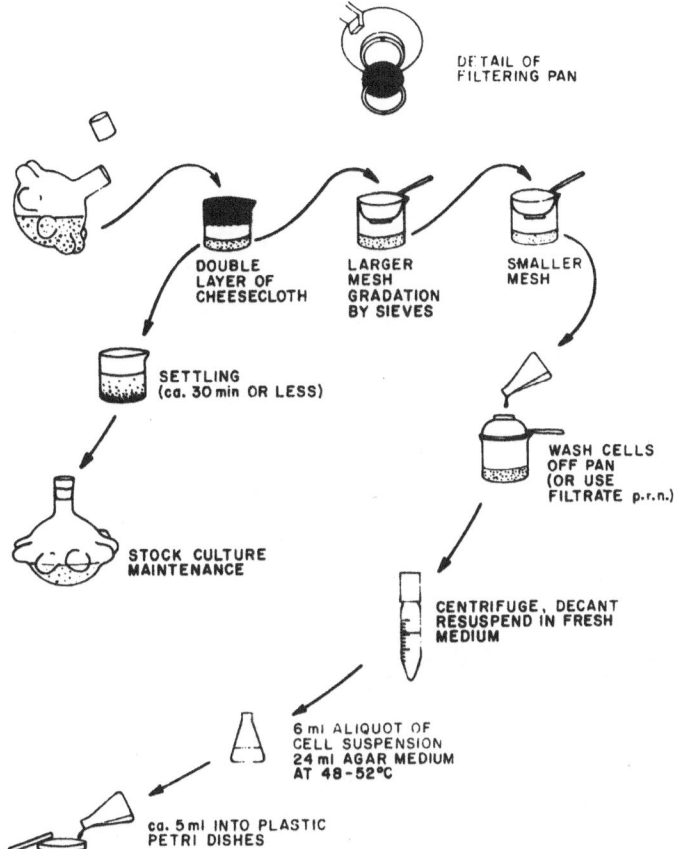

DETAIL OF
FILTERING PAN

DOUBLE
LAYER OF
CHEESECLOTH

LARGER
MESH
GRADATION
BY SIEVES

SMALLER
MESH

SETTLING
(ca. 30 min OR LESS)

WASH CELLS
OFF PAN
(OR USE
FILTRATE p.r.n.)

STOCK CULTURE
MAINTENANCE

CENTRIFUGE, DECANT
RESUSPEND IN FRESH
MEDIUM

6 ml ALIQUOT OF
CELL SUSPENSION
24 ml AGAR MEDIUM
AT 48-52°C

ca. 5 ml INTO PLASTIC
PETRI DISHES

Fig. 1. Diagrammatic representation of procedures followed in filtration of cultured cells grown in suspension to uniform unit sizes for experimentation such as plating in an agar medium. In this illustration, a so-called "nipple flask" [24] has been used to culture the cells; an Erlenmeyer flask serves equally well. Similarly, materials grown on semi-solid media may be suspended in nutrient medium and subjected to separation procedures. p.r.n. means 'as needed'.

Culture of somatic embryos

The somatic embryos are now ready for experimentation in various media and cultures vessels, or for biochemical analysis etc. A typical experiment in semi-solid medium in small plastic disposable Petri dishes may be performed by mixing 6 ml of somatic embryos taken from a centrifuge tube with 24 ml of medium to which autoclaved agar-containing medium has been added. This is done by thoroughly mixing the embryos in the centrifuge tube by swirling and withdrawing 6 ml by pipette. The 6 ml are then introduced into 24 ml of agar medium (the concentration of agar to be used will have to be determined empirically!) which has been cooled to approximately 45 °C. (As noted above, B_{MS} supplemented with "vitamins", and sucrose will sustain the continued growth of these somatic embryos.) Other additions may be tested as desired. The agar must be swirled gently but thoroughly to ensure complete mixing; care must be taken to exclude air bubbles during mixing and final distribution of the tube contents into the final culture vessel. If Falcon # 1006 Petri dishes or an equivalent are to be used, 5 ml per dish is an appropriate amount. It is important to obtain uniform distribution of embryos in the dish. This may be done by pouring the agar-embryo mixture so that approximately two-thirds of the Petri dish is covered; then the dish is quickly tilted so that the remainder of the bottom of the dish is covered. With practice this results in an even distribution of somatic embryos. The snap-tight lids are then attached, and the agar is allowed to harden. Once the agar has hardened, the dishes can be stacked and manipulated.

A count of "X" somatic embryos per one ml in the centrifuge tube is equivalent to "X" embryos per dish in agar medium. Diluting 6 ml of somatic embryos with 24 ml agar medium is equivalent to each ml in the centrifuge tube being diluted 5-fold. Since 5 ml of agar-somatic embryo medium are poured into each Petri dish, each dish contains one ml of the centrifuged embryo suspension. This procedure, even though time-consuming, permits somatic embryos at varying stages of development to be analyzed, grown and manipulated or further evaluated.

Embryogenic suspensions may be passed, successively, through sieves with smaller pore size. Fractions that pass through a # 500 screen (pore size < 25 μm) can be grown but may require extra supplements and care.

A summary evaluation of some key observations and problems of somatic embryogenesis using cell cultures of carrot or other umbellifers according to the "classic" or "conventional model" system described above is outlined briefly as follows:

(a) in the first instance, there has been (and for the foreseeable future will remain) the outstanding question as to where the embryogenic cells come from [25]. One is interested to know whether cells which can develop into somatic embryos are induced to be embryogenically responsive by a specific *in vitro* protocol or regimen, or whether cells which for one reason or another are already embryogenically competent and therefore need only be selected for *in vitro* via permissive media, and cell populations or stocks built up from these for further manipulation. The point here is that in the first instance one is dealing with induction, followed by further manipulation of the circumstances or conditions of expression; in the second, it is a matter of selecting predetermined embryogenic cells or units, building up stocks or populations and manipulating these. While an understanding of the modulation of expression of later stages in somatic embryogenesis is important, it is perhaps a more fundamental question to ask what makes cells embryogenically competent in the first place? Or, conversely, what prevents somatic cells from expressing their presumed ability to become somatic embryos?

Whatever the correct answer to the above question may be, a number of points have emerged from study of somatic carrot embryogenesis in the now "classic" system.

(b) The parallelism with the zygotic embryos is closer if:

(i) their development stems from very small units.(A culture that is heterogeneous as to size fosters the normal development of its smaller ones; a culture rendered uniformly composed of small units by sieving profits by the addition to it of some medium that has been "conditioned" by the growth in it of a large heterogeneous crops of somatic embryos.)

(ii) if the total osmotic concentration of the medium (achieved e.g. by sorbitol) is high and comparable with fluids of embryo sacs which may be in the order of 12–14 atmospheres. (This keeps the somatic embryos small and favors organized growth in contrast to random proliferation.)

(c) the origin and development of small, globular, proembryonic units of carrot in suspension cultures is then fostered by:

(i) a period of dark induction after the cell cultures have been grown and multiplied in continuous light.

(ii) the use, in the ambient medium, of some liquid "pre-conditioned" by the growth of somatic embryos in densely inoculated cultures heterogeneous as to their unit size.

(iii) the duration of the dark induction and the degree of "conditioning" of the medium for somatic embryo formation interact and so promote the embryonic development of the smallest clusters (174 μm which consist of small free cells and cell clusters).

(iv) by withholding reduced carbon (sucrose), while maintaining high osmotic values by use of sorbitol or an equivalent osmoticum in the medium, proembryonic cell cultures of carrot may be maintained in a "poised" state (i.e. their further development is arrested) until these conditions are reversed.

(v) the cells of the pro-embryonic globules that develop under these conditions into somatic embryos are small, with a few small vacuoles and a cytoplasm which is notably free of unusual inclusions but is rich in microtubules [24, 25].

However, despite all the above it has been very clear for a long time that much more needs to be done to prescribe fully the growth factor requirements and the external conditions which are most conducive to somatic embryogenesis under *in vitro* conditions [3, 7].

Some recent alternatives or modifications to the "classic" embryogenic carrot system protocols

It has long been suspected that the mineral element composition of culture media influences virtually every aspect of *in vitro culture* including initiation, maintenance and development of somatic embryos. Most often though, these suspicions have been ignored and emphasis has been placed on the extent of somatic embryo formation, and this only when exogenously added growth regulators were used to initiate and maintain the cultures. Also, considerable attention has been given to developing protocols that can yield somatic embryos, which in turn, develop into mature plants phenotypically identical to the plant from which the original explant was derived. Moreover, the requirements for, or the ability to respond to, certain mineral elements and other media components has widely been thought to be controlled by the exogenously added growth regulator(s) and that these requirements then direct the developmental fate of those cells. Hence, there has been little incentive to investigate inorganic components in isolation from the more usual organic, hormonal additives.

The procedures to be described below show that the classic or model carrot system, as described above, generally thought in the past only to be achievable in a reliable way via added growth regulator(s), is controllable with high confidence without the use of any added growth regulator(s).

Hormone-free (H-F) procedures

Using mericarps

1. Surface sterilize "seeds" of a carrot cultivar such as "Scarlet Nantes" (see seed sterilization procedures 1 or 2 above on p. 4 and following) and place them directly onto semi-solid or in liquid medium (see Fig. 2 on p. 25 for a schematic representation of the responses in cultures to be described).
   All tests are carried out in 15 mm × 100 mm plastic Petri dishes containing 40 ml of various media made semi-solid by the addition of 0.7% agar or in 125 ml Erlenmeyer flasks containing 25 ml of medium.
2. Inoculate onto each dish either five "seeds" (number 1 in the scheme) or five "seeds" each devoid of their zygotic embryo after germination (number 3) but still retaining their endosperm. (Mericarps rendered free of their embryos by dissection are a poor tissue source but will work). Culture conditions may include either continuous darkness, or light at 22 °C [18]. Under these conditions, radicles emerge at approximately day 5 (number 2).
3. Physically separate the seedling (number 4) from the "seed" after day 7–10 (number 3). At day 7–10, a small non-proliferating callus is observable on nearly all fruits at the site of embryo exit if the mericarp remains on the culture medium (number 3a). Proembryos are observable as small 0.5 mm to 1.0 mm diameter, whitish globules at day 30–35 (number 3b).

At this time the proembryos can be left undisturbed (3b to 3c) or can easily be separated from the mericarp (3b′-5 or 3b″-3h″) with the same results. As illustrated in Fig. 2 (3b′–5), normal development from the proembryo (3b′) through the heart stage (3c′), torpedo stage (3d′) stage, cotyledonary stage (3e′) and the plantlet is possible on certain basal media.

Figure 2 (3b″-3h″) also illustrates that normal development up to the torpedo stage (3d″) can result in abnormal cotyledonary stage embryos (3e^{III}-3^{VI}) which in turn can give rise to a secondary somatic embryogenesis system. Proembryos, depicted in 3f″, arise from the axis of pre-existing cotyledonary stage embryos. The secondary proembryos give rise to both somatic embryos and adventitious shoots (3g″) that again give rise to more proembryos (3h″) in a cyclical manner (3h″-3b″).

For example, certain media foster embryo growth that follows pathway 3b″-3e^{V} thus resulting in "neomorphs", i.e. embryonic forms which do not go on to further developmental stages unless they are cultured on an appropriate basal medium prior to the torpedo stage of their development (horizontal arrows 3c″ to 3c′).

The mineral element composition of the basal medium quantitatively affects zygotic embryo "germination", and the number of mericarps giving rise to proembryos. It furthermore qualitatively affects the size of the proembryos, morphology of the cotyledonary stage embryos and the continued growth of embryos into plantlets or the initiation of cyclical secondary embryo formation.

Table 2 provides a base or skeleton formulation of salts to which nitrate or ammonium may be added for experimentation. In this laboratory, the base formulation is referred to as "DS-5a salts". For tissue maintenance as PGSPs, add 1 mM NH_4Cl (54 mg/l) and buffer with 25 mM MES (free acid) (5000 mg/l) titrated with potassium hydroxide to pH 4 to 4.2. Alternatively, add 10 mg/l of casein hydrolysate (acid hydrolyzed). For continuation of the embryogenic process beyond the PGSP stage, add 5 mM NH_4Cl (270 mg/l) or 5 mM NH_4OH and buffer with 25 mM $MES.H_2O$ free acid (5000 mg/l) at pH 6.0. The former may be referred to as "maintenance medium" and the latter to as "embryo growth medium". Further points concerning other additions to the medium such as activated charcoal will be covered below.

Table 2. DS-5a "Salts" or Base Formulation. Note: nitrogen source and buffering to be carried out as needed. (See text for details).

	mM	mg/l
$Ca(H_2PO_4).H_2O$	1.0	252.10
$MgSO_4.7H_2O$	1.0	246.49
NaCl	0.5	29.23
H_3BO_3	0.1	6.18
$MnSO_4.H_2O$	0.1	16.9
$ZnSO_4.H_2O$	0.01	2.88
KI	0.005	0.83
$Na_2MoO_4.2H_2O$	0.0001	0.024
$CuSO_4.5H_2O$	0.0001	0.025
$CoCl_2.6H_2O$	0.0001	0.024
$Na_2FeEDTA.2H_2O$ } $FeSO_4.7H_2O$ }	0.1	*
thiamine HCl	0.0037	1.0
sucrose	58.4	20,000.00

* 27.8 mg/l $FeSO_4.7H_2O$; 37.23 mg/l $Na_2EDTA.2H_2O$ or 33.6 m/l Na_2EDTA.
See [15 and 16] for details of chelated iron solutions.

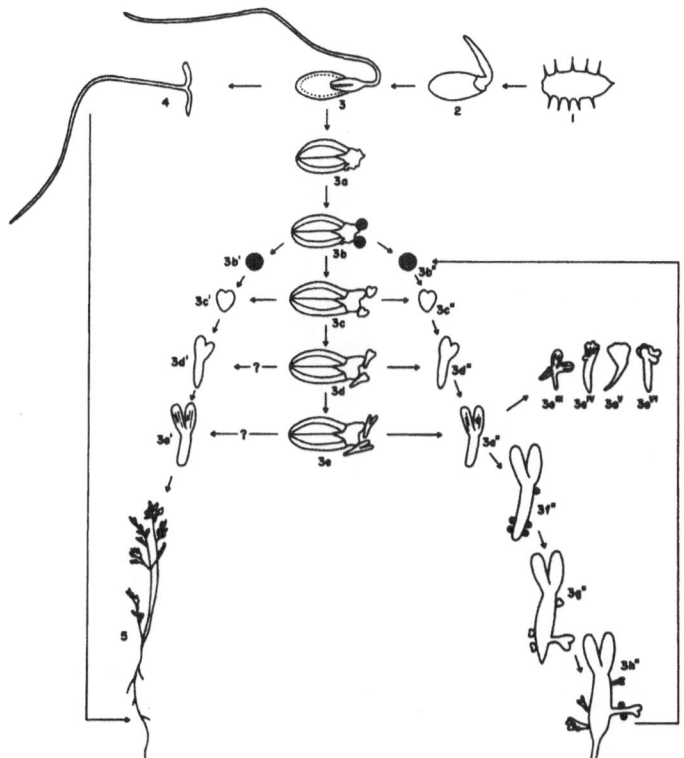

Fig. 2. Somatic embryogenesis of carrot from mericarp tissue in the absence of exogenous hormone. See text for details. The two question marks indicate that these steps have not been investigated although there is no reason to believe that they would not proceed in a predictable fashion.

Using zygotic embryos

Early work suggested that mericarps were a dramatically better source of embryogenically competent cells than zygotic embryos. Moreover, only wounded zygotic embryos were able to yield somatic embryos and only a few at that [18].

1. Attempts to increase somatic embryo production from wounded zygotic embryos were realized, however. Experiments showed that H-F nutrient medium, containing 1 mM NH_4^+ as the sole nitrogen source, fostered production of somatic embryos from wounded zygotic embryos to the same high degree found with mericarps. Intact embryos or seedling hypocotyls, on the other hand, virtually never produce somatic embryos [19].
2. Somatic embryo formation from wounded zygotic embryos occurs mainly from the cotyledons, but root tips and hypocotyls (also with shoot tips present) of mature zygotic embryos are responsive as well [19].
3. On media containing unreduced nitrogen, somatic embryo formation leads to the generation of vigorous cultures entirely comprising somatic embryos at various stages of development which in turn proliferate still other somatic embryos [19].
4. Growth on nutrient medium containing 1 mM NH_4^+ prevents the initially formed somatic embryos from developing into later stages, but does not prevent cell multiplication. This apparent inability to continue development results in cultures consisting entirely of preglobular stage proembryos.
5. This medium containing 1 mM NH_4^+ as the sole nitrogen source apparently does not induce somatic proembryo formation. Instead, the medium is best thought of as permissive to the expression of certain embryogenically determined cells within zygotic embryos, especially those in the cotyledon [20].
6. Histological examination has confirmed that preglobular stage proembryos from wounded zygotic embryos are maintained on 1 mM NH_4^+-containing medium.
7. The first-formed somatic embryos continue development into later embryo stages, without continued secondary embryo proliferation, if the medium pH is maintained above 4.5. (Tested at pH 4.5, 5.0, 5.5 and 6.0) [21].
8. The establishment of cultures consisting entirely of preglobular stage proembryos is a process, not an event. The first-formed somatic embryos multiply in the beginning as globular stage embryos, only when the pH of the medium is allowed to fall during the culture period. During each successive culture period, the volume per tissue mass made up of preglobular stage proembryos increases. A total of 4 to 6 transfers of the entire tissue mass after initiation of somatic embryos is required to establish a culture consisting of preglobular stage proembryos.
9. Establishment of preglobular stage proembryos can be hastened by repeated mashing or wounding of the first-formed globular stage embryos at the time transfers are made.

Some specifics in the protocols

Maintaining preglobular stage proembryos on semi-solid medium

1. Preglobular stage proembryos are maintained and multiplied on semi-solid hormone-free (H-F) medium initially at pH 4.3 (cf. Table 2, p. 24 with an appropriate nitrogen source) by subculturing at 2–3 week intervals. To assure a good crop of preglobular stage proembryos, subcultures should be carried out every 2 weeks. Each subculture is carried out by moving about one-half of the tissue mass, so that four tissue masses, each of approximately 5 mm, diameter are cultured per 100×15 mm diameter plastic Petri dish. Each dish contains between 40–50 ml of medium.
2. At the end of a 3 week growth period, the masses referred to above should have grown to about 1 cm diameter. If all the tissue from one dish is collected and dispersed in liquid, there should be about 1 ml settled cell volume. Settled cell volume is determined by allowing the cells to settle for 10 min in either 30 ml or 15 ml plastic pre-sterilized graduated tubes. The tubes are then tapped gently on a surface 3 times before measuring the volume.

Continuation of somatic embryogenesis from PGSPs into later stages

1. PGSPs larger than 140 μm in diameter often yield multiple or twinned somatic embryos. PGSPs should therefore be sieved through a # 100 sieve (140 μm pore size) and collected in a beaker (cf. Fig. 1, p. 17). This can then be poured through a # 200 sieve (74 μm pore opening), and the cell clusters which are retained are backwashed onto the # 200 sieve. Thus, one is using cell clusters with a diameter between 74 and 140 μm. This yields a relatively uniform population of units and hence the response to experimental treatments will be more uniform.
2. PGSPs so collected may be washed once with "embryo growth" medium but this is not absolutely necessary.
3. The tradeoff in the procedure that seeks to assure a more uniform response and single somatic embryos is, however, that there will be a considerable reduction in the usable volume of PGSPs.
   This can be offset, of course, by maintaining larger stocks of PGSPs in readiness for experimentation.
4. For experiments in either liquid or on semi-solid medium, the final settled cell volume per volume of medium should be around 0.1 ml settled cell volume to 50 ml medium. To enable subsequent calculation of the percent progression of distributed or suspended units into later embryonic stages, an aliquot of the cell clusters should be removed and counted in a haemocytometer or other appropriate cell counting device. For example, if you have 0.5 ml settled cell volume, adjust the total volume to 5 ml; then pipette 1 ml out for each 50 ml culture vessel. For semi-solid medium,

drop-by-drop application and spreading out of the PGSPs over the solidified surface by some appropriate means needs to be carried out. In the case of liquid medium, inoculation is carried out directly by means of a pipette.

Semi-solid medium

A mixture of washed agar (National Formulary of the U.S.A. grade has been very satisfactory), 0.4% w/v plus 0.1% gellan gum (Kelco Gelrite) permits an appropriate level of gelation for semi-solid medium preparation for "maintenance medium". For "embryo growth medium" 1.2% washed agar is satisfactory. See [19] for agar washing procedures.

The use of activated charcoal

Activated charcoal provides a major benefit in the progression of PGSPs to later embryo stages [20, 21]. Powdered activated charcoal, maintained and stored so that activation level is assured [9], can be used around 1% w/v for liquid cultures or 0.5% for semi-solid agar media. These percentages are approximate figures, however, since different charcoals will behave differently and will have to be tested in a given setting [9, 21] to ascertain what the response level generated will be. The objective will necessarily determine what is used and how. Having determined an appropriate kind of charcoal to use, the matter becomes one of assuring even distribution. Various procedures are used by different laboratories to assure even distribution of particles and to prevent settling before gelling occurs. One convenient way is to pour semi-solid Petri dishes in two or three layers, allowing each layer to solidify before the next is added. The final or "finishing layer" should be the smallest in volume.

Charcoal-impregnated filter papers (# 508, Schleicher and Schuell, Keene, New Hampshire), placed on a semi-solid surface of a Petri dish have been used successfully [20, 21] and provide the added advantage of enabling the handling of tissues conveniently since tissues resting on the papers can be transferred and manipulated with relative ease.

These papers have to be washed, however, and autoclaved twice to assure sterility [21].

Use of low external pH to replace synthetic auxin like 2,4-D in maintaining and multiplying embryogenic cells of carrot

Integration of the knowledge gained from the use of some of the classic methods described above to initiate embryogenic cultures of carrot have led to the view that cultures that contain embryogenic cells generated by any means can be, in time, sustained virtually exclusively at the pre-globular proembryogenic level by exposing them to low pH in medium devoid of the initiating/maintaining auxin. This procedure may be likened to a "jump-start" wherein a hormone like

2,4-D is used to initiate an embryogenic culture (see p. 9 and following, above), but low pH is then employed to clean-up the culture, i.e. enrich it with embryogenic cells, and to maintain it predominantly in the PGSP-replicative condition [22]. This approach would seem to have major advantages since numerous plant parts do not yield demonstrably embryogenic cultures in media devoid of growth regulator like 2,4-D. In the carrot system, for example, seedling hypocotyl does not yield embryogenic cultures like explanted wounded zygotic embryos, or mericarps [19]. Therefore, depending on one's objectives, one can initiate embryogenic cultures using hormone or whatever means possible, and then transferral to hormone-free [H-F] medium at low pH may be effected. In this way, the advantages of both procedures may be capitalized upon.

Ultimately, one cannot emphasize too strongly that objectives should determine strategies adopted and that somatic embryo-generation procedures should be closely tied to the experiments in question.

Acknowledgments

The feasibility of the investigations referred to here arose from long continued support from the National Aeronautics and Space Administration. This help is gratefully acknowledged.

References

1. Ammirato PV (1983) Embryogenesis. In: Evans DA, Sharp WR, Ammirato PV, Yamada Y (eds) Handbook of Plant Cell Culture. Vol I, Techniques for propagation and breeding, pp 82–123. New York: Macmillan.
2. Ammirato, PV (1984) Induction, maintenance, and manipulation of development in embryogenic cell suspension cultures. In: Vasil, IK (ed) Cell Culture and Somatic Cell Genetics of Plants. Vol. I, Laboratory procedures and their applications, pp 139–151. Orlando: Academic Press.
3. Carman J (1990) Embryogenic cells in plant tissue cultures: Occurrence and behavior. *In Vitro* Cellular and Developmental Biology 26: 746–753.
4. Choi JH, Sung ZR (1989) Induction, commitment, and progression of plant embryogenesis. In: Kung S-d, Arntzen CJ (eds) Plant Biotechnology, pp 141–159. Boston, London: Butterworths.
5. Fitter MS, Krikorian AD (1982) Plant Protoplasts. Some guidelines for their preparation and manipulation in culture.San Diego: Behring Diagnostics.
6. Halperin W (1966) Alternative morphogenetic events in cell suspensions. American Journal of Botany 53: 443–453.
7. Komamine A, Matsumoto M, Tsukahara M, Fujiwara A, Kawahara R, Ito M, Smith J, Nomura K, and Fujimura T (1990) Mechanisms of somatic embryogenesis in cell cultures-Physiology, biochemistry and molecular biology. In: Nijkamp, HJJ, van der Plas, LHW, van Aartrijk, J (eds) Progress in Plant Cellular and Molecular Biology, pp 307–313. Dordrecht and Boston: Kluwer Academic Pubs.
8. Krikorian AD (1982) Cloning higher plants from aseptically cultured cells. Biological Reviews 57: 151–218.
9. Krikorian AD (1988) Plant tissue culture: Perceptions and realities. Proceedings of the Indian Academy of Sciences (Plant Science) 98: 425–464.
10. Krikorian AD (1989) Introduction to: Growth and organized development of cultured cells by Steward, FC, Mapes, MO, Mears, K, Amer. J. Bot. 45: 705–708.1958. In: Janick, J (ed) Classical Papers in Horticultural Science, pp 40–55. W.H. Freeman, New York.
11. Krikorian AD, Kelly K, Smith, DL (1987) Hormones in plant tissue culture and propagation. In: Davies PJ (ed) Plant Hormones and their Role in Plant Growth and development. pp. 592–613. Martinus Nijhoff/Dr. W. Junk, Dordrecht, Netherlands.
12. Murashige T, Skoog F (1962) A revised medium for rapid growth and bioassays with tobacco tissue. Physiologia Plantarum 15: 473–497.
13. Nomura K, Komamine A (1986) Somatic embryogenesis in cultured carrot cells. Development, Growth and Differentiation 28: 511–517.
14. Scott DR, Walz, AJ, Manis, HC (1966) The effect of Lygus spp. on carrot seed production in Idaho (Hemiptera: Miridae). University of Idaho Research Bulletin no. 69: 1–12.
15. Singh M Krikorian AD (1980) Chelated iron in culture media. Annals of Botany 46: 807–809.
16. Singh, M, Krikorian AD (1980) White's standard nutrient solution. Annals of Botany 47: 133–139.
17. Small E (1978) A numerical taxonomic analysis of the *Daucus carota* complex. Canadian Journal of Botany 56: 248–276.
18. Smith DL, Krikorian AD (1988) Production of somatic embryos from carrot tissues in hormone-free medium. Plant Science 58: 103–110.
19. Smith DL, Krikorian AD (1989) Release of somatic embryogenic potential from excised zygotic embryos of carrot and maintenance of proembryonic cultures in hormone-free medium. American Journal of Botany 76: 1832–1843.
20. Smith DL, Krikorian AD (1990) Somatic proembryo production from excised, wounded zygotic carrot embryos on hormone-free medium: evaluation of the effects of pH, ethylene and activated charcoal. Plant Cell Reports 9: 34–37.
21. Smith DL, Krikorian AD (1990) Somatic embryogenesis of carrot in hormone-free medium: External pH control over morphogenesis. American Journal of Botany 77: 1634–1647.

22. Smith DL, Krikorian AD (1990) Low external pH replaces 2,4-D in maintaining and multi-plying 2,4-D-initiated embryogenic cells of carrot. Physiologia Plantarum 80: 329–336.
23. Steward, FC, Krikorian AD (1971) Plants, Chemicals and Growth. Academic Press, New York.
24. Steward FC, Israel HW, Mott RL, Wilson, HJ, Krikorian AD (1975) Observations on growth and morphogenesis in cultured cells of carrot (*Daucus carota* L.). Philosophical Transactions of the Royal Society of London B 273: 33–53.
25. Street HE (1976) Experimental embryogenesis- The totipotency of cultured plant cells. In: Graham CF, Wareing, PF (eds) The Developmental Biology of Plants and Animals, pp 73–91. Philadelphia: W.B. Saunders.
26. Waris H (1959) Neomorphosis in seed plants induced by amino acids. I. *Oenanthe aquatica.* Physiologia Plantarum 12: 753–766.

Plant Tissue Culture Manual **A10**: 1–28, 1992.
© 1992 *Kluwer Academic Publishers.*

Low density cultures: microdroplets and single cell nurse cultures

G. SPANGENBERG[1] & H.-U. KOOP[2]
[1] *Institute for Plant Sciences, Swiss Federal Institute of Technology, CH-8092 Zürich, Switzerland*
[2] *Institute of Botany, University of Munich, D-8000 Munich 19, FRG*

Introduction

The development of low density cell culture systems is mainly based on two assumptions. Firstly, although plant cell (protoplast) cultures are commonly referred to as relatively homogeneous cell populations the results of closer analysis have often shown the existence of a high degree of cellular hetero-geneity. Thus, plant cell cultures do not consist of a large number of uniform cells, but rather are a population of cells showing a range of variation in individual genotype, phenotype and age [5]. This range of variation can be further increased by the use of mutagenic agents, genomic mixing and recombi-nation achieved through protoplast fusion and genetic transformation. Secondly, most chemically defined plant cell culture media will not support cell division at low cell densities, commonly less than about 9,000–15,000 cells/ml: the *critical inoculum density* [41], and plant cell culture under optimal conditions deals with population densities in the range of 10^4–10^6 cells/ml.

Thus, the need for the development of methods allowing "density indepen-dent" plant cell growth, in order to improve the possibility of recovering rare variants, becomes evident.

In this context, the use of conditioned media [14, 25, 2], nurse cell layers [4, 10] or growing cells in very low culture volumes in multiwell dishes or hanging droplets [15, 11, 6, 26] have all been successfully used for growing clones from single cells or protoplasts.

In addition, all these techniques have been applied for *in vitro* selection thus dissecting the "natural" heterogeneity of the original population of plant cells in more homogeneous subpopulations [25, 10], or for the "density indepen-dent" culture of the extended range of – more or less rare – variants produced by mutagenic agents [44], by protoplast fusion [12, 22, 27] and through genetic transformation [6].

An extension of the techniques for culturing small numbers of cells, following the strategy of reducing the volume of culture medium, namely using *micro-cultures* [13, 43, 15, 11] was achieved with the success of *ab initio* microculture in chemically defined and unconditioned culture medium of individually selected single plant cells in *nanodroplets* (10–50 nl) of culture medium pro-tected from evaporation by a layer or droplets of mineral oil [16, 18, 36]. This was based on a reduction of the volume of culture medium to a similar cell population density as used in mass culture, i.e., if a population density in the range of 10^4–10^5 cells/ml is required for *en masse* plant cell or protoplast culture, then *one single* cell or protoplast should be cultured in a nanodroplet of 100–10 nl (microculture or microdroplet culture), respectively.

A similar development for establishing techniques in order to culture small numbers of "target" cells following the principle of "nurse culture" [24, 7, 30, 42] led finally to a method for the nurse culture of individual cells [8, 31].

Our aim is not to provide a detailed review of tissue culture methods dealing with the culture of small populations of plant cells, but rather to emphasise two experimental protocols for the individual culture of single "target" cells, namely 1) the individual selection and exclusive culture of defined single plant cells or protoplasts in nanodroplets of unconditioned culture medium injected into separate microdroplets of mineral oil: *microculture* or *microdroplet culture*, and 2) the individual culture of "target" plant cells grown close to, but physically separated from a second plant cell population (feeder cells), supporting their growth: *single cell nurse culture*.

We also describe a computerized hydraulic system which allows a) the microscopical selection of defined plant cells from a population and their individual inoculation, applied in both experimental protocols outlined above, b) the preparation of the *microculture chamber* in the case of the microdroplet culture and c) the positioning of micro-tools used in different micro-manipulation steps, e.g. positioning of microelectrodes for one-to-one micro-(electro)fusion of preselected pairs of protoplasts, or positioning of micro-injection needles and holding capillaries.

A. Microdroplet Culture

Individual cell culture represents *a priori* a useful tool for *in vitro* selection of different cell types and studies on their physiology, for the analysis of cell differentiation programs and cell-to-cell interactions, as well as a culture cloning step after the performance of genetic manipulations at the single cell level: *single cell engineering* [32, 20, 38].

By using a microculture chamber [18] and a computer-aided instrumental setup [32, 20] the microscopical selection of single plant cells or protoplasts and their individual culture at the same population density as in mass culture can be performed.

This microdroplet culture system has been so far successfully used for the individual selection, and culture in microculture chambers, of mesophyll proto-plasts of *Nicotiana tabacum* [18], hypocotyl protoplasts of *Brassica napus* [36] and protonemata protoplasts of *Funaria hygrometrica* [23] and *Physcomitrella patens* [1] and microspores of barley [3], as well as for the analysis of conditioning effects at the single cell level in different cell types of *B. napus* [33].

Single cell engineering applications, based in part on this micromanipulation instrumental setup or on the culture in microdroplets of the micromanipulated plant cells or protoplasts, have been reported for a number of systems, such as: intranuclear microinjection of protoplasts and karyoplasts of *B. napus* [37], microfusion of preselected pairs of protoplasts of *N. tabacum* [17, 19], cell reconstitution by protoplast-subprotoplast microfusion in *B. napus* [35]; protoplast microfusion in *F. hygrometrica* [35], *in vitro* fertilization via microfusion of isolated gamete cells in *Zea mays* [23]; defined somatic hybridization via protoplast microfusion in *N. tabacum* [39]; organelle transfer [9] and defined cybridization via protoplast-cytoplast microfusion in *N. tabacum* [40].

Procedures

1. Major Equipment: Instrumental Setup for Microdroplet Culture

As already mentioned, the microdroplet culture is mainly based on a microprocessor-aided instrumental setup (Fig. 1) [32, 20].

This instrumental setup which is assembled in an air-flow cabinet in order to fulfill the sterility requirements, consists basically of an optical and operational-positioning unit (a); a mineral oil filled hydraulic system (b and c) and a control unit (d) (Fig. 1A, B). In more detail, these units are:

a) An inverted microscope (IM 35 Zeiss microscope with Nomarski optics, C. Zeiss, Oberkochen, FRG) with a microprocessor-controlled programmable microscope stage (EK8b-S4 microscope stage and control unit MCC 13 JS RS232, Gebr. Märzhäuser OHG, Wetzlar, FRG, or from Lang-Elektronik, D-6338 Hüttenberg, FRG) driven by adjustable stepmotors in x-, y- and z-axis. This allows the positioning of a microculture chamber and a selection chamber while selecting single defined cells under microscopical control (optionally based on a TV-screen image) from the population of cells kept in the selection chamber, and their individual inoculation into microdroplets of the microculture chamber (see section 3) (Fig. 1A-C).

   Alternatively, the insert of the microscope stage can be replaced by a number of different inserts allowing for example the positioning of up to six coverslips for the automatic preparation of microculture chambers (see section 2.3) or the use of standard culture dishes (Petri dishes, slide chambers, microtiter plates) in various combinations.

b) A hand-pulled selection microcapillary, fixed to a holding device driven by the z-axis stepmotor, is connected via a teflon tubing filled with mineral oil to:

c) A microprocessor-controlled nanoliter-pump (modified diluter Microlab M, Hamilton, Bonaduz, Switzerland), which can be programmed for volume, speed and direction of the pumping, thus allowing delivery and withdrawal of volumes in the nanoliter-range (Fig. 1E) and, finally:

d) A microprocessor (Apple II, Apollo-IIASKF/64K, AAA Electronic GmbH, Freiburg, FRG with interface CCS 7710, California Computer System, CA 94086, USA or from Pete & Pam Micro Computers, Haslingden, Rossendale, Lancs BB4 5HU, U.K.; and IBS-AP2, IBS Computertechnik, Bielefeld, FRG) controlling the positioning electronics of the x-, y- and z-axis stepmotors of the programmable microscope stage as well as the nanoliter-pump (Fig. 1B). Alternatively other computers can be used. For software see section 2.3.8.

   With these elements, this instrumental setup can be programmed to perform the different working steps required, for example:

A) *Fully-automatic preparation of a microculture chamber:* The setup adjusts under the optical axis of the inverted microscope to a given position in x- and y-axis, and the microcapillary supported in the z-axis-device is lowered so that the tip of the microcapillary contacts the coverslip, placed on the microscope stage at the pre-fixed position. The nanoliter-pump is activated to deliver a pre-established volume with a given speed, and after delivery of the microdroplet the microcapillary is lifted and adjusted to the next position following a pre-determined array in x- and y-axis. Thus all working steps required for the preparation of microculture chambers for microdroplet culture can be performed with minimal intervention by the experimenter.

B) *Selection of defined cells and their transfer into microdroplets:* While activating the same functions, the position of a particular cell in the selection chamber can be chosen under microscopical control by the experimenter with the aid of a joy-stick, manually controlling the positioning system in x-, y- and z-axis. The selection microcapillary is lowered over the chosen cell and the selected cell withdrawn in a volume in the nl-range after foot-switch activation of the nanoliter-pump. Then the selection microcapillary is automatically lifted, the microscope stage moved so that a microdroplet at a pre-established position of the microculture chamber is now under optical control, the microcapillary is lowered into the nanodroplet of culture medium covered by a microdroplet of mineral oil and by a second activation of the nanoliter-pump, the chosen cell can be delivered into the desired microdroplet.

If access to all of these major pieces of equipment is not possible, successful microdroplet culture, namely preparation of microculture chambers as well as selection of defined plant cells into microdroplets, can still be performed, but obviously with more intervention and expertise required from the experimenter.

In this case, the positioning system in x-, y- and z-axis, namely the automatic microscope stage with the three stepmotors and the corresponding control-electronics can be omitted and replaced by a standard manually moveable micro-scope stage (simply adapted for two 24 × 40 mm coverslips, namely for the selection and microculture chambers) (Fig. 2C).

In addition, the selection microcapillary will then also be manually controlled while delivering the nanodroplets of culture medium into the oil microdroplets (see section 2.3) and selecting cells from the selection chamber under microscopical control and individually transferring them into previously (also mainly manually) prepared microculture chambers (see section 3). For the exclusive application in microdroplet culture, this "less automatic" version of the instrumental setup [18]

Fig. 1. Major equipment for single cell culture.
A) Setup for microprocessor-controlled selection and transfer into microculture of defined plant cells.
B) Components of the selection and microculture setup: (1) inverted microscope, (2) nanoliter-pump, (3) microprocessor and (4) 1 μl microdroplets-dispensing device.
C) Microprocessor-controlled microscope stage with x-, y-, z-stepmotors (1, 2 and 3) with support for microtools (4) e.g. for microelectrodes for microfusion (5).
D) Detailed view of 1 μl microdroplets-dispensing device from B(4) showing tubing connection to a selection microcapillary.
E) Detailed view of microprocessor-controlled nanoliter-pump from B(2).

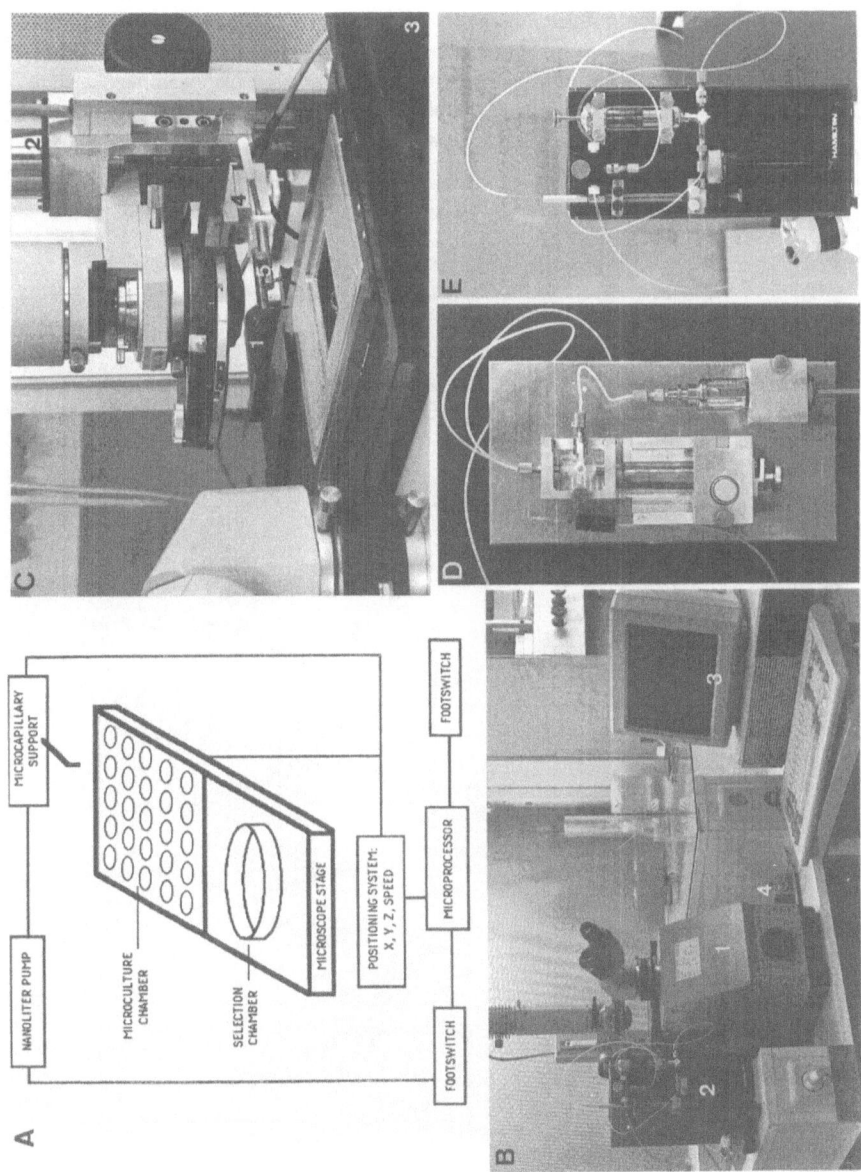

still allows a skillful experimenter – after a reasonable training period – a comparably fast (e.g. selection and transfer of 200 cells/h) and sometimes more flexible performance. Thus, for microdroplet culture alone, in the authors' experience, this simplified version is equally suited compared to the full computer-aided setup. However, if more precise positioning of micro-tools for particular micromanipulation purposes is envisaged, e.g. positioning of micro-electrodes for micro(electro)fusion, then the full automatic setup will provide a more accurate and efficient working basis.

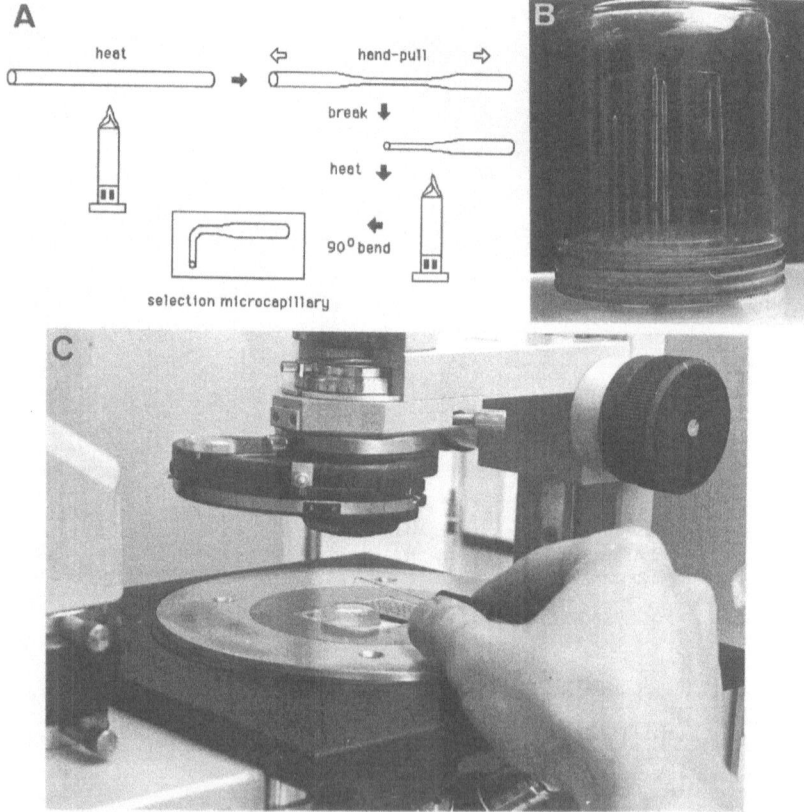

Fig. 2. Selection microcapillary and manual selection of defined plant cells.
A) Preparation of selection microcapillaries for single cell culture.
B) Ready-to-use drawn selection microcapillaries.
C) Manual selection of defined plant cells from selection chamber into microdroplets.

2. Preparation of Selection Microcapillaries, Selection Chambers and Microculture Chambers

In the following sections a detailed description of the different components and tools required as accessories for performing microdroplet culture is presented; important steps and practical considerations regarding the procedure and minor equipment are stated in the notes.

2.1. Preparation and maintenance of selection microcapillaries

The selection microcapillaries are hand-drawn from 50 µl disposable micropipettes (Blaubrand, 50 µl, green color code, No. 708733, Brand, IntraMark) over a small flame.

Steps in the procedure

1) Heat the center of the micropipette until soft, remove it from the heat, wait 1–2 s and then quickly pull the two ends apart, so that the 127 mm long micropipettes are elongated by an approximately 3–4 cm pulled central region (Fig. 2A).
2) Break the pulled micropipettes with the aid of a diamond pencil in the middle of the narrowed portion and bevel the rough broken tips with fine polishing-paper.
3) Bend the drawn-out portions of both extended capillary-halves to a right angle by holding the capillary-half over a small flame and heat just before the expanded portion and wait until the drawn-out portion bends down due to gravity (Fig. 2A, B).

Notes

2.1.1 It is recommended to pull micropipettes with a series of different opening diameters. On average, for selection of cells the inner diameter of the open tip should be approximately 2-3 fold the diameter of the cell to be selected, and thus will vary if selection of large vacuolated protoplasts, karyoplasts, cytoplasts, gamete cells, etc. is intended.

2.1.2 For the delivery of liquids (e.g. mineral oil; 2.5 M sucrose or culture medium), the choice of the appropriate microcapillary will depend mainly on the viscosity and volume of the solution to be handled. For viscous liquids (e.g. mineral oil and 2.5 M sucrose), which are normally used for the delivery of 1 µl microdroplets (see section 2.3), larger opening-capillaries (inner diameter: 600–800 µm) are preferred. For the delivery of nanodroplets (e.g. 10–20 nl) of standard plant cell culture media into mineral oil microdroplets, narrower opening-capillaries (inner diameter: 300–500 µm) are recommended.

2.1.3 Drawn-out selection microcapillaries should be carefully checked (eventually, requiring the use of a stereomicroscope), keeping in mind the following criteria: a) a correct bend at 90°, b) 10–15 mm length of the bent terminal portion, c) 45–60 mm length of the not-drawn portion, d) evenly broken, smooth and thin-walled tips. After strictly discarding those not fulfilling the requirements from a) to d), they can be sorted in two or three categories depending on the opening size of their tips for different experimental purposes (see 2.1.1 and 2.1.2) and finally, autoclaved (Fig. 2B).

2.1.4 Pulled selection microcapillaries can be re-used as long as they fulfill the sterility and other requirements (see notes 2.1.1–2.1.3). The chances for re-use in subsequent experiments are increased if the same capillary is used for the same experimental purpose (e.g. a capillary appropriate for delivering mineral oil droplets should be used exclusively for dispensing oil and not sucrose solution). As long as the microcapillary is sterile, it can be replaced (and kept under axenic conditions) if, with the same delivery-setup, another experimental purpose (and thus another microcapillary, e.g. for selecting protoplasts, is required) is to be accomplished. If the microcapillaries do not work properly any longer, it is recommended to discard them, instead of attempting a time-consuming, cumbersome washing procedure.

2.2 Preparation of selection chambers

Steps in the procedure

1) Prepare selection chambers by glueing a polycarbonate ring (5 mm high, 24 mm outer diameter) onto a 24 × 40 mm coverslip (coverglass AL No. 9161040, Menzel, FRG) with silicone glue (Fugendichtung Praktikus, Praktikus-Chemie, D-4048 Grevenbroich 5, FRG) (Fig. 2C). The rings are made out of 5 mm high transverse sections from a 24 mm diameter polycarbonate cylinder.
2) Bevel the sections to a ring leaving inside an empty volume with the shape of an inverted truncated cone (13 mm inner lower diameter, 22 mm inner upper diameter).
3) Smear the wider ring-base with the silicone glue and smoothly press onto the center of the coverslip and slightly rotate while pressing, in order to allow for an even distribution of the glue and a tight sealing-contact between the poly-carbonate ring and the coverslip.

Notes

2.2.1 After hardening of the glue, the selection chambers can be autoclaved. Care should be taken to allow possible cytotoxic solvents (e.g. acetic acid) to evaporate from the silicone glue, so a 2–3 days air drying of the selection chambers is recommended before autoclaving.

2.2.2 Other types of selection chambers are conceivable, however, they should allow for good optical quality (e.g. recommended coverglass), should be easily autoclavable (e.g. polycarbonate, autoclavable silicone glue), should have dimensions providing 1–2 ml content and should be compatible with following considerations: manipulation of reasonable populations of plant cells at a time, with population densities compatible with easy identification of "target" cells to be selected and facilitating the withdrawal of individual cells in volumes of 10–50 nl; provide an easy access to the "target" cell with the selection microcapillary considering a working distance of 3–4 cm between the microscope stage and the condensor (Zeiss IV Z/7(0,63) with removed front lens) of the inverted microscope; have a convenient surface area/volume ratio under comparatively high evaporation conditions (e.g. air-flow cabinet during 1–3 h), etc. Standard poylstyrene Petri dishes can also be used as selection chambers, where optimal optical conditions are not critical.

2.2.3 Non-leaking selection chambers can be re-used (after washing with a tissue culture compatible detergent and re-autoclaving). Otherwise, the polycarbonate ring can be removed and newly glued onto another coverslip.

2.2.4 When preparing selection chambers, a small amount of silicone glue remains at the inner edge of the ring on the glass surface. This remaining silicone can be helpful for removing mineral oil, eventually clogging the tip of the selection microcapillary, while transferring cells into microdroplet culture (see section 3).

2.3 Preparation of microculture chambers

Steps in the procedure

1) Prepare the standard microculture chambers [18, 32] from 24 × 40 mm cover-slips (microcover glasses Thomas No. 6663-F82, A. Thomas Co., USA, marked e.g. by cutting off one corner), where 50 1.0 µl microdroplets of 2.5 M sucrose are positioned onto the coverslip in an array of five rows with 10 microdroplets per row (Fig. 3A, B).
2) Immerse the coverslips carrying the sucrose-droplets for 1–2 s in a 2% dimethyl-dichlorosilane solution in 1,1,1 trichloro-ethane (Repel Silane, No. 1850-252, LKB, Sweden), drain and leave for 10–15 min. in a fume-hood to evaporate the solvent.
3) Wash with warm tap-water until the sucrose-droplets are completely removed, rinse with ethanol and double distilled water and carefully dry with optical-quality paper.
4) Each coverslip is then placed, with the side where the sucrose droplets had been positioned facing upward, in a two-compartment 6 cm diameter culture dish (organ tissue culture dish Falcon, No. 3037, Becton-Dickinson & Co., Cockeys-ville, USA) and UV-sterilized for 15–20 mins. The coverslips are now handled under sterile conditions for positioning 50 1,0 µl droplets of autoclaved mineral oil (paraffin oil for spectroscopy Merck, No. 7161, Merck, Darmstadt, FRG) exactly in the center of the non-siliconized circular areas previously occupied by the sucrose-droplets.
5) Finally with the aid of the fully automatic instrumental setup, inject nanodroplets of culture medium (normally in the range of 15–100 nl) with the microcapillary centered in each mineral oil droplet previously placed on the coverslip (Fig. 3C).
6) The prepared microculture chamber is transferred back to the two-compartment culture dish, 1.5–2 ml of autoclaved 0.2 M mannitol are pipetted into the outer compartment of the culture dish, thus serving as moist chamber, and finally the latter is sealed with Parafilm (American Can Company, Greenwich, CT 06830, USA) (Fig. 3B).

Notes

Notes 2.3.1 to 2.3.8 deal with minor equipment required for and some technical aspects concerning the preparation of microculture chambers, notes 2.3.9 to 2.3.14 refer to the different components of the microculture chamber *per se*.

2.3.1 The 50 1.0 µl droplets of sucrose and of mineral oil are distributed routinely, while using the full automatic setup with the microprocessor controlled nanoliter-pump (see section 1), 3,350 µm apart from each other in rows of 10 with a distance of 3,400 µm between the rows, when the standard microculture chamber is prepared. However, the volume and number of droplets and their distribution in x- and y-axis as well as the distance between droplets and rows can be freely chosen and programmed.

2.3.2 If no programmable microscope stage/positioning system is available (see section 1), still a very accurate distribution of the 1.0 µl microdroplets (of sucrose solution or mineral oil) can be achieved by manually applying them onto the coverslip after placing it on a plate with a dotted array (5 rows per 10 dots each, every dot marked approximately 3 mm from each other on a mm-squared paper). The application of the 1.0 μl droplets of sucrose solution or mineral oil can then be performed with the aid of a simplified dispenser-device (described under note 2.3.3 and illustrated on Fig. 1D).

2.3.3 This simplified 1.0 μl microdroplets-dispensing arrangement consists of a 10 ml syringe (luer-lok, Ultra-Asept for Braun-Perfusor, No. 872 855/0, B. Braun Melsungen AG, D-3508 Melsungen, FRG) used as oil reservoir, a 50-teeth mechanical device (dispensor-device Hamilton PB 600-1, No. 83700, Hamilton Bonaduz AG, CH-7402 Bonaduz, Switzerland) driving the piston of a 50 μl syringe (Hamilton type 705N, No. 80500 Hamilton) and allowing the stepwise (max. 50 steps, 1 μl each) emptying of its contents, and the connections (B. Braun Melsungen, FRG) required to link via a three-way-cock (Edwards-Bentley model K75B, Bentley Lab. Europe, 5404 AA Uden, Holland) both syringes to the selection microcapillary (see description on section 2.1). All mechanical elements are anchored to a stable plate and all required connections are achieved by a clear flexible teflon tubing (Tefzel tubing 1.8 mm outer diameter; 0.8 mm inner diameter, No. 19-7435-01, Pharmacia, Sweden). The 10 ml syringe oil-reservoir is attached to one of the openings of the three-way cock to provide auxiliary negative pressure (e.g. for loading the selection microcapillary and in part the tubing system with sucrose solution or mineral oil) or positive pressure (e.g. while pulling out the piston of the 50 μl syringe linked to the 50-teeth mechanical device). The terminal connection of the teflon tubing to the selection microcapillary is achieved by a plastic screw-type fitting (tubing connector for 1.8 mm o.d. tubing, No. 19-7476-01, Pharmacia, Sweden), thus allowing an easy exchange as well as tight fitting of selection microcapillaries, as well as representing a holding element in the case of manual selection. The entire arrangement functions as an hydraulic system filled with autoclaved mineral oil (Fig. 1D).

2.3.4 The positioning of the culture medium nanodroplets into each of the 50 1.0 μl mineral oil microdroplets loaded onto the coverslip in a specified array is done under optical control (the use of a 2.5 × objective is recommended, as it allows for the observation of four mineral oil microdroplets on the image field and thus provides an easy orientation in x- and y-axis). The delivery of the nanodroplets of culture medium is performed with the aid of the nanoliter-pump (mentioned in section 1), independently of whether the PC-controlled microscope stage/positioning system or a manual control of the positioning of the microscope stage and selection microcapillary is used. The modifications performed on the commercially-available dispenser (Microlab M, Hamilton, Bonaduz) for adapting it to the purpose of microdroplet culture are described under note 2.3.5 and illustrated on Fig. 1E.

2.3.5 The nanoliter-pump for microdroplet culture is a slightly modified dilutor MircolabM (Hamilton, Switzerland). On the front plate of the commercially available dispenser a 1 μl or, alternatively, a 5 μl syringe (Hamilton, type 70001N, No. 80100, and 7005N, No. 200305, respectively, Hamilton Bonaduz) is installed and mechanically linked to the movable dispensing axis-device driven by a stepmotor (with 1000 steps working range) of the dilutor. Thus, per step minimal volumes of 1 nl (1 μl/1000 steps) or 5 nl (5 μl/1000 steps), respectively, can be delivered or withdrawn with this arrangement in the range of 1 nl up to 5 μl (by varying the number of driven steps). In addition, the arrangement consists of a 10 ml syringe (luer-lok, Ultra-Asept for Braun-Perfusor, No. 872 855/0, B. Braun Melsungen AG, D-3508 Melsungen, FRG) connected to the selection microcapillary and to the 1 μl (or 5 μl) syringe via a three-way cock and the required clear flexible teflon tubing for linking all these components. Further modifications (not shown) can include three outlet valves and tubings for separate delivery of sucrose, oil and culture medium, respectively and a separation unit in the sucrose outlet to prevent mixing of oil and sucrose in the tubing system. An oil reservoir is attached to one of the openings of the central three-way valve of the nanoliter-pump to provide oil supply while delivering (from the 1 μl or 5 μl syringes) volumes in excess to the total syringe content (1 or 5 μl, respectively). The 10 ml syringe, also anchored to the front plate of the nanoliter pump, acts as a wider-volume range reservoir providing auxiliary negative pressure (e.g. for loading the selection microcapillary and in part the teflon tubing with e.g. culture medium) or positive pressure (e.g. for flushing the hydraulic system free of disturbing air-bubbles). The entire arrangement acts as an hydraulic system and is filled with autoclaved mineral oil (Fig. 1E).

2.3.6 The ready-to-use microculture chamber should already contain a fraction of the final volume of culture medium/microdroplet which will, by the end, after individual transfer of the selected cell into the microdroplet (see section 3) complete the expected volume for achieving an optimal volume of culture medium/cell ratio (e.g. nearly equivalent to an optimal population density in *en masse* culture). Routinely, at this initial step, volumes in the range of 10—30 nl of the culture medium are injected in each oil microdroplet and directly dispersed onto the unsiliconized area of the coverslip forming a flat central nanodroplet of culture medium, evenly covered by the oil microdroplet (Fig. 3C-D).

2.3.7 Independently of the mode of positioning control in x- and y-axis (via the PC-controlled microscope stage or a manually slidable standard stage) as well as for the z-axis (manually holding the selection microcapillary at the screw-type fitting or fixing it to a support which can be vertically displaced by a PC-controlled stepmotor anchored under the condensor of the inverted microscope), the reproducible delivery of defined nanodroplets of culture medium into the oil microdroplets can only be achieved under microscopical control (see note 2.3.4) with accuracy using the nanoliter-pump (described under note 2.3.5). In the case of the PC-controlled positioning system, by pressing once the corresponding foot-switch the first position (center of an oil microdroplet) is automatically placed under the optical axis and the z-axis stepmotor vertically driving the selection microcapillary lowered (and a final adjustment by the experimenter of the coordinates in all axis via a joystick is allowed). These finally adjusted coordinates can then be saved by re-pressing the same foot-switch. By pressing then once the foot-switch corresponding to the nanoliter-pump control, the programmed volume (e.g. 10—30 nl) is delivered at the first chosen position. From now on, alternating the positioning system and the nanoliter-pump will be automatically activated repeating the delivery of culture medium into the oil-microdroplets following the pre-programmed positioning pattern (freely selectable from a corresponding menu) as well as a selectable time factor (for allowing the reproducible delivery of different volumes or of different culture media with distinct viscosities into the oil microdroplets) in a stand-alone function without further intervention by the experimenter. This procedure for preparation of microculture chambers can be performed on parallel with up to six coverslips on the programmable microscope stage. In the case of the manually driven standard microscope stage and manual holding of the selection microcapillary, a single foot-switch activating the nanoliter-pump has to be pressed by the experimenter every time the tip of the microcapillary is immersed into each oil microdroplet and serially completing the 50 positions on a standard microculture chamber. This procedure can be performed with e.g. two coverslips placed on the standard microscope stage at a time.

2.3.8 Required software for both systems outlined before (see note 2.3.7): a) exclusive control of the nanoliter-pump, and b) coordinated control of positioning system and nanoliter-pump, allowing a flexible arrangement of delivered and withdrawn volumes and of positions and arrays of droplets can be obtained upon request for Apple II and IBM-compatible PCs from the authors. For IBM-PCs XT or AT (or compatibles) equipped with a standard games part and two serial parts, extensive software (developed by the second author) is available. In addition, similar software for Apple II, IBM and Epson systems for the same experimental purposes can be obtained upon request from E. Kranz (Department of Botany, University of Hamburg, D-2000 Hamburg, FRG), U. Lagercranz (Department of Plant Breeding, Swedish University of Agricultural Sciences, S-750 07 Uppsala, Sweden) and F.L. Olsen (Department of Physiology, Carlsberg Laboratory, DK-2500 Copenhagen Valby, Denmark), respectively.

2.3.9 The rationale behind the preparation of the microculture chamber [18] is to protect the nanodroplets of culture medium from evaporation by covering them with oil microdroplets separately (in order to guarantee "individual" culture) positioned onto a coverslip (in order to achieve good optical quality). To avoid merging of the oil microdroplets on the surface of the coverslip, the latter is siliconized between the oil droplets, which is achieved as indicated on Fig. 3.

2.3.10 For defining the array of unsiliconized circular areas on the coverslip (Fig. 3A) a viscous solution, namely 2.5 M sucrose (filter-sterilized, not autoclaved: otherwise there is a loss of viscosity) is recommended over the 2.0 M described in the original protocol [18]. In order to be able to discriminate between the 2.5 M sucrose solution and the oil (contained in the hydraulic system used for dispensing 1 µl droplets) it is recommended to include in the sucrose solution a vital dye (e.g. cochenille red or food colorant E124). While delivering the sucrose droplets, care should be taken to avoid oil coming out of the microcapillary, as it would be dissolved by the Repel-Silane solvent, the glass surface below the sucrose droplet would then be partially siliconized and thus the nanodroplet of culture medium will not be properly dispersed onto the coverslip.

2.3.11 It is recommended for the preparation of the oil microdroplets and for filling the hydraulic system to use good quality paraffin oil (.e.g. paraffin oil for spectroscopy, Merck No. 7161, E. Merck, Darmstadt, FRG). For the microdroplets preparation, some activated charcoal (activated charcoal Merck No. 2514, particle size: 1.5 mm) for adsorbing possible cytotoxic contaminants from the oil, can be added to the oil. The oil can then be saturated with water (by adding double distilled water to the oil-activated charcoal suspension and autoclaving it; re-autoclave it at least once a week). The oil may have to be discarded after multiple re-autoclaving.

2.3.12 The use of standard coverslips (e.g. Menzel AL No. 9161040, 24 × 40 mm, type 1, Menzel, FRG) has proven to be as good as more expensive microcover glasses of better — lead free — quality (e.g. microcover glasses Thomas No. 6663-F82, A. Thomas Co., USA) for most experimental purposes. However, if problems are encountered in microculture of particular "sensitive" plant cell types, attention should be paid to quality requirements of all the different components of the microculture chamber, as cells cultured in these "micro-environments" may react more sensitively than if mass-cultured.

2.3.13 Immediately after preparation of the microculture chambers (after the nanodroplets of culture medium have been injected into the oil microdroplets), these should be kept before use in a moist chamber (two-compartment Falcon dish, as described in section 2.3). Only freshly prepared micro-culture chambers (not more than 2—4 days after injection of the culture medium nanodroplets) should be used. Coverslips with sucrose droplets (but not siliconized) can be prepared well in advance and be stored (it is even recommended to let them air-dry for 2—3 days before siliconization if freshly prepared 2.5 M sucrose solution is used). Microculture chambers in preparation should not be stored after the oil microdroplets have been positioned onto the coverslip (this makes the delivery of the nanodroplets of culture medium more difficult).

2.3.14 Regarding the culture medium to be used for microculture, the medium used for mass culture of the cells or protoplasts of interest has — in general — proven to be satisfactory. It is however recommended, before starting microculture experiments, to optimize the composition of the culture medium and culture conditions (macro- and micronutrients, hormones, vitamins, cell density, buffering system, pH, osmolality, etc.) using a population of protoplasts (e.g. using the MDA-technique; [29]) until achieving a reproducible and efficient plant cell (protoplast) mass culture protocol. If so, optimization of the microculture conditions would be restricted to eventually checking particular parameters affecting the performance of the individually cultured cells (e.g. volume of nanodroplets), however, in general, no "microdroplet culture-specific" requirements to the culture medium are necessary.

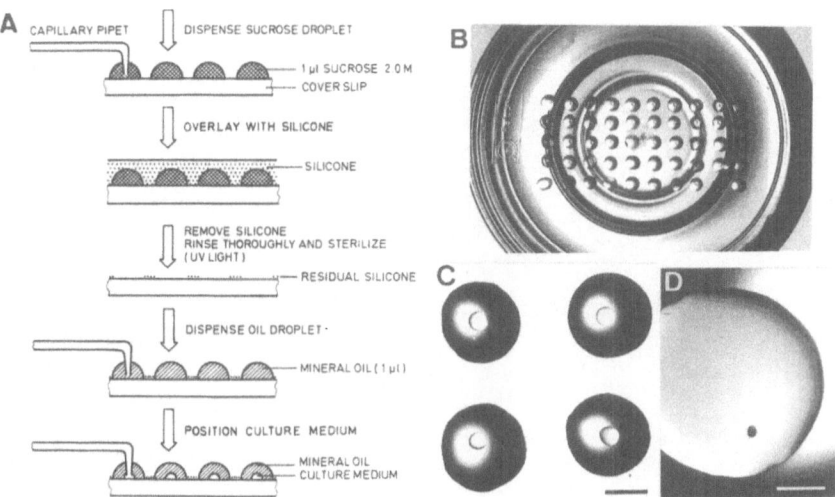

Fig. 3. Microculture chamber for individual culture of defined plant cells.

A) Preparation of microculture chambers for single cell microdroplet culture.

B) Microculture chamber in a moist chamber (two-compartment dish).

C) Detailed view of part of a microculture chamber. Each 1 µl droplet of mineral oil contains 30 nl of culture medium. Bar: 1 mm.

D) Individually selected mesophyll protoplast of *N. tabacum* in a nanodroplet of culture medium. Bar: 500: µm.

3. Selection of defined Plant Cells and Individual Culture in Microdroplets

As mentioned before, the instrumental setup for microdroplet culture (see section 1, Fig. 1) in both its variations (namely, manual positioning or full automatic control) has been used by the authors for the selection of defined protoplasts out of heterogeneous cell populations, their transfer into microdroplets on microculture chambers, their individual culture and further growth until plant regeneration for isolated protoplats and/or fusion products of following plant species: *N. tabacum* [18, 19, 39, 40], *B. napus* [34, 32], *F. hygrometrica* [23] and *P. patens* [1]. Details on the performance of protoplasts from the above mentioned species are provided in the original references. Only some relevant aspects, for achieving successful microculture of protoplasts from these species are described in the following, illustrating some considerations for microdroplet culture of more general interest when dealing with other species. Some special recommendations are presented in the notes.

Once the technical elements of the instrumental setup for microdroplet culture, the microculture chambers and other accessories are established, the reader interested in performing microdroplet culture of protoplasts or plant cells from a system efficiently working in his own hands at a mass culture scale, will be faced first with two questions: 1) which initial volume should the nanodroplet of culture medium have? and 2) which type of cell or protoplast should be selected?
   Indeed these two parameters play an essential role on the success of the individual microculture of defined protoplasts (establishment of cell clones, finally leading to plant regeneration) using the experimental setup described here.

As regards to the initial volume of the nanodroplet of culture medium into which a single protoplast has been selected, volumes in the range of 30–50 nl have proven adequate for tobacco mesophyll protoplasts and rapeseed hypocotyl protoplasts. Larger nanodroplets of 125 nl culture medium have proven advantageous for the microdroplet culture of protonemata protoplasts of *F. hygrometrica*.
   Concerning which type of cell should be selected, obviously, the experimental purpose followed, e.g. cell cloning for *in vitro* cell line selection for increased yields of secondary metabolites, or microisolation of heteroplasmic cells after protoplast mass-fusion, etc., would be a key factor upon this decision.

However, the experimenter interested in microdroplet culture will necessarily identify discrete morphologically distinguishable cell (or protoplast) types (e.g. according to cell size, distribution and number of organelles, degree of vacuolization, vacuolar accumulation of anthocyanins, evident presence of cytoplasmic strands, etc.) in the apparently homogeneous population of cells one is dealing with, and thus learn to look at populations of plant cells as collections of individuals. In this context, it is recommended to invest particular time and attention in comparing the behaviour (e.g. viability, rates of cell division and microcallusing, etc.) in microdroplet culture of selected subpopulations of discrete individual-types.

In addition, since the experimental setup described allows on average the manipulation of 100 to 200 cells/hour, the experimenter is restricted to the use of subpopulations limited in number selected out from plant cells (or protoplasts) populations which are heterogeneous. This inherent variation as well as the variability present between different protoplast preparations, due at least in part to a series of external factors affecting protoplast culture [28], may account for much of the variation observed in different experiments while performing microdroplet culture with apparently the "same" protoplast system. Nevertheless, in the authors' experience, plant regeneration from microcultured protoplasts of *F. hygrometrica* can be reproducibly achieved with overall frequencies in the range of 40–60% [23], for tobacco protoplasts on average between 3–60% [18, 39, 40] and for rapeseed hypocotyl protoplasts with frequencies of 0.1–0.2% [32, 38]. However, cell division frequencies from microcultured protoplatsts can significantly vary — within the ranges described — from one microculture chamber to another.

For achieving sustained cell division in microculture it is in some cases (e.g. for tobacco and rapeseed protoplasts) required to add fresh culture medium to the microdroplets. Optimization of a "feeding"-system for microcultured protoplasts is recommended, as this factor may even prove to be a critical step [36]. The optimal "feeding" schedule should be determined considering following factors: a) starting time for the addition of fresh culture medium (normally not required before first cell divisions have been observed), b) frequency of the addition of culture medium (e.g. every 3–4 days has proven convenient for tobacco and rapeseed protoplasts), c) volume of the added culture medium and dilution rate (e.g. 10–30 nl in the case of tobacco protoplasts, and exclusively in those microdroplets where protoplasts have divided, in order to avoid dilution of the population density of microcultured protoplasts being "laggards" in cell division), d) composition, e.g. regarding hormones and osmoticum used, of the added culture medium.

In this way, microcolonies (consisting of ca. 20–100 cells) derived from *ab initio* microcultured protoplasts can be obtained under oil microdroplets, after 20–30 days of individual culture, with frequencies of up to 90% (on average, however, 20–30%) of the cases for tobacco and between 1–2% for rapeseed hypocotyl protoplasts. In the case of *F. hygrometrica* protoplasts, protonema filaments will be formed from the individually selected protoplasts after 10–15 days in individual culture without requiring addition of fresh culture medium to the microdroplets containing individual protoplasts in nanodroplets of initially 125 nl culture medium (having a size which will directly allow — with the aid of the tip of a needle — to transfer them out of the oil microdroplets directly to standard culture conditions for further growth and plant regeneration (see: [23]) in up to 60% of the cases.

In a species-dependent manner, the timepoint for transferring the microcultured protoplast-derived microcolonies out of the oil microdroplets and further culture steps have proven to be critical in particular cases. In the case of tobacco protoplasts — after 20–30 days in microculture — microcalluses can be individually transferred onto wells of multiwell dishes containing 1 ml of agarose-solidified morphogenesis medium and efficiently regenerated to plants. However, for microcultured rapeseed hypocotyl protoplasts, the microcolonies obtained — after up to 1 month in micro-culture — will necessarily require for further growth intermediate culture steps, including first, the transfer onto 1 µl wells of polycarbonate microdishes containing 500 nl of liquid culture medium and later, up to three additional microculture steps with a progressive increase of the culture volume (2—5 fold from one step to the next) per microcallus for finally establishing callus clones with potential for plant regeneration [32].

Notes

3.1 In most cases, conditions optimized for microdroplet culture for one particular genotype of one species allow for acceptable performance (e.g. viability, cell division, microcallus formation and finally plant regeneration) of protoplasts from other genotypes of the same species or intraspecific protoplast microfusants. This holds true: a) for tobacco protoplast microculture in initial 55 nl droplets of culture medium PNT [39] for *N. tabacum* cv. Xanthi, cv. Petit Havana SR 1, cv. Badischer Burley, tobacco cms analogs with cytoplasms of *N. debneyi*, *N. bigelovii* and *N. suaveolens* and some of their pairwise fusants; b) for rapeseed hypocotyl protoplast microculture in initial 60 nl droplets of culture medium PBN-7 [34, 36, 32], for *B. napus* cv. Bronowski, cv. Tower and rapeseed cms analog with cytoplasm of *Raphanus sativus* and some of their pairwise fusants; and c) for protonemata protoplasts of the moss *F. hygrometrica* in microdroplets of initial 125 nl of culture medium Fha [23] for wild type and four different auxin-resistant development mutants and some of their pairwise microfusants. In most cases tested, if genotype-dependent differences in behaviour in microdroplet culture were evident, a similar trend was detectable in mass culture. Nevertheless, genotype dependent "microdroplet culture specific" differences cannot be generally ruled out.

3.2 If no significant differences in performance of individually selected protoplasts in microculture can be detected in a particular range of initial volume of the nanodroplets of culture medium, e.g. for *F. hygrometrica* protoplasts in the volume range of 30—125 nl, then the larger volumes are recom-mended, as: a) a faster delivery of the protoplasts is possible during the selection and transfer into microculture, b) further "feeding"-steps, where fresh culture medium is added/exchanged to the microdroplets, can be limited to a minimum or fully avoided, thus reducing tedious working steps, risk of contamination, osmotic shock of the microcultured cells, etc.

3.3 When transferring a defined cell from the selection chamber to a microdroplet of the microculture chamber the optimal total initial volume of the nanodroplet of culture medium will be divided into a portion already present under the oil microdroplet in the ready-to-use microculture chamber and the residual volume required for withdrawing the cell from the selection chamber. A halving of the desired total volume of the nanodroplet of culture medium for fulfilling both purposes is recommended, if possible. In general, the portion of the microdroplet volume of culture medium already present under the oil microdroplet (*recipient volume*) should not be less than 5—10 nl (in order to avoid "burying" the selected cell in oil while transferring it) — if possible — and the same holds true for the withdrawing volume from the selection chamber (*transfer volume*) while selecting the cell (in order to facilitate its delivery into microculture, as minor elasticities in the mineral oil-filled hydraulic system cannot be completely excluded). If larger volumes (more than 10—30 nl) are tolerable as initial total nanodroplet volume, then the recipient volume can be increased and the transfer volume can be split into two working steps of the nanoliter pump: one for withdrawing only culture medium (without the cell) from the selection chamber and the second for withdrawing the cell to be selected and transferred. In this way, the cell to be transferred will be more terminal to the tip of the selection microcapillary and thus rapidly delivered into the microdroplet while activating the nanoliter-pump and a risk of selecting the "target" cell with an "undesired" cell from the selection chamber — due to large negative pressure and withdrawing volume — will be reduced.

3.4 Indicated overall plant regeneration frequencies for the microdroplet culture of protoplasts of tobacco, rapeseed and *F. hygrometrica* are calculated as % of the individually selected protoplasts transferred into microdroplets which formed microcolonies under microculture conditions, these latter being transferred out of the oil microdroplets and separately plated onto further culture steps and finally regenerating to plants.

B. Single cell nurse culture

The feasibility, in principle, to culture individual higher plant cells by exploiting the supportive effect of "nurse cells" was initially demonstrated by Muir *et al.* [24] in their classical paper, in which it was shown that a higher plant cell line can be derived from an individual cell. There, an individual cell was placed on a sheet of filter paper on top of a nurse callus serving as a physical but not a diffusion barrier between the individual and the nurse cells. Here, we describe the simultaneous nurse culture of many single cells in a culture system, which uses an agarose layer containing a number of pits for inoculation with pre-selected single cells and a stainless steel basket with a nylon sieve or dialysis membrane bottom as a container for nurse cells in liquid medium [8]. With this culture system [8] a number of physiological parameters relevant to single cell nurse culture have been analysed [31].

In contrast to nanodroplets single cell nurse culture does not provide completely defined culture conditions. However, the system may offer a number of advantages, which can be summarized as follows:

A) *Transfer of single cells*
Cells are routinely cultured in a total volume of 4 ml (see below) of culture medium. Thus, it is not critical to use a defined and very small volume for withdrawal and transfer of single cells from selection chambers. Therefore, a nanoliter pump is not mandatory for this procedure, although still highly recommendable for selection and transfer of individual cells.

B) *Renewal of culture medium*
Culture medium is simultaneously replenished for a whole set of target cells using standard pipettes. Thus it is not necessary to "feed" fresh medium into individual microculture chambers as might be the case for nanodroplets. Single cell nurse culture therefore provides a culture system for individual cells which is much easier to handle.

C) *Same culture dish during regeneration from single cells or protoplasts to macroscopically visible colonies*
Whereas specific "feeding" protocols (compare section 3) and a precise step-up procedure regarding the total culture volume at each developmental stage of a microcolony may be required when using nanodroplets, in single cell nurse culture, the cell and the colonies derived thereof are maintained at the same sites in the same culture dish until macrocolonies of 1 to 2 mm diameter have been achieved, sufficient in size for transfer to standard culture conditions. This again contributes to the ease of handling of this procedure.

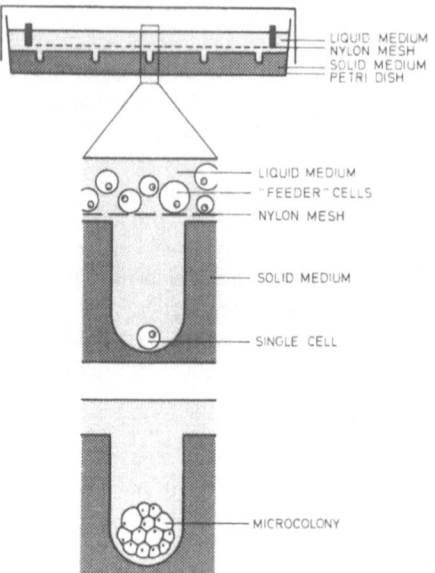

LIQUID MEDIUM
NYLON MESH
SOLID MEDIUM
PETRI DISH

LIQUID MEDIUM
"FEEDER" CELLS
NYLON MESH

SOLID MEDIUM

SINGLE CELL

MICROCOLONY

Fig. 4. Schematical representation of the single cell nurse culture system. The upper part shows the total setup, a section with a single protoplast and the feeder layer is depicted in the central part, the lower part demonstrates a single microcolony after removal of the feeder layer.

Procedures

5. Major equipment

Basically, the same hardware and software as described for microdroplet culture is used for selection and transfer of individual cells in single cell nurse culture. Since this culture system uses however standard Petri-dishes (3.5 cm diameter) rather than coverslips, a stage insert containing at least one position for a Petri-dish is required.

6. Making a single cell nurse culture dish

A special casting tool (Fig. 5B) is required for preparing a single cell nurse culture dish. It is machined by inserting the 25 pins (diameter 1 mm, lengt 7 mm) of a standard computer plug (Fig. 5A, male DB 25P, ITT Cannon, Santa Ana, CA, USA) in five rows of five pins into a brass holder. The distance between pins in a row and between rows is 4 mm. The brass holder is adjusted to provide a distance of 0.5 mm between the lower ends of the pins and the inner surface of the bottom of the Petri-dish, when the holder is inserted into the Petri-dish (Fig. 5C). The casting tool has a handle for easier insertion into and removal from the Petri-dishes. It is sterilized by autoclaving prior to its usage. For casting of single cell nurse culture dishes, 2 ml of liquified 2% agarose (LMP agarose, BRL, Bethesda) made up with the respective culture medium, autoclaved and stored in a water bath at 45 °C, are pipetted into a Petri-dish, and the casting tool is inserted. After 10 to 15 min at room temperature, 1 ml of liquid culture medium is added on top of the solidified agarose layer, and the casting device is carefully removed.

Notes

6.1 The dimensions of the system are critical: if the agarose layer is thinner, the pits are not deep enough, and the slightest turbulence will remove the single cells from the pits.

6.2 Single cell nurse culture dishes can be stored in a refrigerator for at least one week. During storage the density of the medium tends to increase causing the cells inoculated into the agarose pits to float rather then sediment into the pit. The liquid layer should then be replaced with fresh culture medium prior to inoculation with single cells.

7. Making a basket for feeder cells

Two stainless steel rings (30 mm outer diameter; 6,5 mm height) are used for preparing a basket for feeder cells (Fig. 5D). The outer diameter of the lower part of the upper ring is adjusted to slightly less than the inner diameter of the lower ring, such that when sliding the lower ring onto the lower part of the upper ring, a dialysis membrane (or nylon sieve) is safely held in place between the two rings. When the bottom of the basket, providing a physical but not a diffusion barrier is attached, the parts of the membrane, protruding between the two rings are trimmed off using small scissors. Whereas nylon sieve of 5 μm pore size is also suitable, a single layer of dialysis tubing (43 mm width, Sigma # 9527), providing much better optical conditions, is recommended [31].

Notes
7.1 Stainless steel rings and barrier membranes should be carefully cleaned by boiling in a suitable laboratory detergent and subsequent rinses with bidistilled water.

7.2 When using dialysis membrane, it should not be allowed to dry during the whole procedure in order to prevent formation of folds.

7.3 Assembled baskets are autoclaved and stored in bidistilled water and excess water is drained off prior to inserting the basket into the Petri dish.

8. Assembling a nurse culture dish after inoculation of the pits in the agarose layer with single cells

After inoculation with single cells, a feeder basket is inserted into the culture dish, carefully avoiding entrapment of air between the agarose layer and the basket, and feeder cells, suspended in one ml of culture medium are pipetted into the basket.

9. Culture procedure

As with culture of individual cells in nanodroplets, protocols for single cell nurse culture have to be adjusted to the type of cells used in a particular experiment. For tobacco leaf protoplasts the following procedure is used:
1) use the culture medium as described earlier [8, 31];
2) adjust the density of nurse cells to 5×10^4 cells per ml of the total volume (4 ml);
3) once a week, and starting after one week of culture, remove one eight to one quarter of the liquid phase outside and inside of the basket and replace with fresh culture medium;
4) gradually reduce the osmolality after approximately three weeks of culture.

Following this procedure, colonies become macroscopically visible (300 µm diameter) after three to four weeks of culture and feeder baskets can be removed at this time. Routinely, about 50% to 90% of the single cells develop into colonies, and subsequently, into plants.

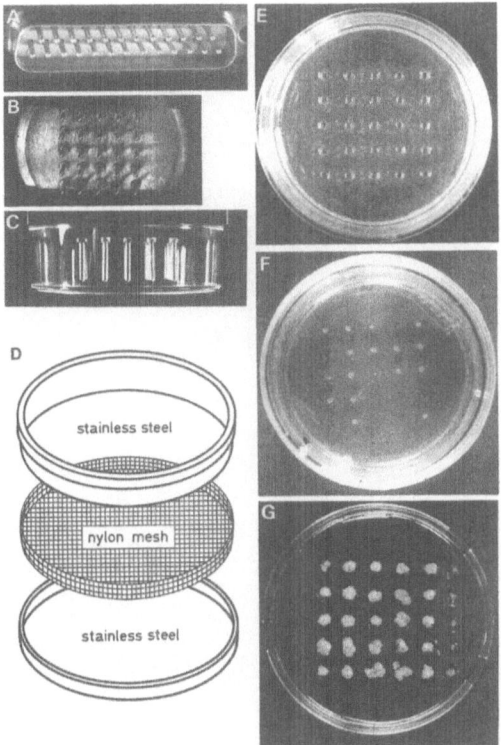

Fig. 5. Single cell nurse culture system.

Fig. 5A-E: preparation of single cell nurse culture dish. Fig. 5F: microcolonies derived from single protoplasts, 5G: test for contamination of single cells with feeder cells during single cell nurse culture.

Fig. 5A: computer plug containing 25 gold plated pins, Fig. 5B: pins removed from the plug and mounted into a brass stamp, Fig. 5C: stamp with pins inserted into the bottom of a standard plastic Petri-dish (3.5 cm diameter) for molding of an agarose layer, Fig. 5D: schematical representation of a custom made basket for taking up of the feeder cells, Fig. 5E: Petri-dish with agarose layer containing pits for inoculation with single cells, Fig. 5F: microcolonies in a single cell nurse culture dish, derived from leaf protoplasts of *N. tabacum* after four weeks of culture and after removal of the feeder layer, Fig. 5G: growth of 25 colonies produced from single cell nurse culture of leaf protoplasts of a kanamycin resistant line of *N. tabacum* and 5 colonies produced from feeder protoplasts derived from a non-resistant wildtype; all the colonies originating from culture of single protoplasts show normal growth on selection medium. Illustrations are not to scale.

References

1. Abel WO, Knebel W, Koop HU, Marienfeld JR, Quader H, Reski R, Schnepf E, Spörlein B (1989) A cytokinin-sensitive mutant of the moss, *Physcomitrella patens*, defective in chloroplast division. Protoplasma 152: 1–13.
2. Bariaud-Fontanel A, Tabata M (1988) Somaclonal variation in the berberine-producing capability of a culture strain of *Thalictrum minus*. Plant Cell Rep 7: 206–209.
3. Bolik M, Koop HU (1991) Identification of embryogenic microscopes of barley (Hordeum vulgare L.) by individual selection and culture and their potential for transformation by microinjection. Protoplamsma (*in press*).
4. Cella R, Galun E (1980) Utilization of irradiated carrot cell suspensions as feeder layer for cultured *Nicotiana* cells and protoplasts. Plant Sci Lett 19: 243–252.
5. Collin HA, Dix PJ (1990) Culture systems and selection procedures. In: Dix P (ed) Plant Cell Line Selection, pp 3–17. Weinheim: VCH Verlagsgesellschaft.
6. Crossway A, Oakes JV, Irivne JM, Ward B, Knauf VC, Shewmaker CK (1986) Integration of foreign DNA following microinjection of tobacco mesophyll protoplasts. Mol Gen Genet 202: 179–185.
7. De Ropp RS (1955) The growth and behaviours in vitro of isolated plant cells. Proc Roy Soc B 144: 86–93.
8. Eigel L, Koop HU (1989) Nurse culture of individual cells: Regeneration of colonies from single protoplasts of *Nicotiana tabacum, Brassica napus* and *Hordeum vulgare*. J Plant Physiol 134: 577–581.
9. Eigel L. Koop HU (1990) Organelle transfer by protoplast-subprotoplast microfusion: transfer of very low numbers of wild type chloroplasts leads to variegated regenerants. Abstract In: Proceedings VII IAPTC Meeting, Amsterdam 1990, pp 209.
10. Ellis BE (1985) Characterization of clonal cultures of *Anchusa officinalis* derived from single cells of known productivity. J Plant Physiol 119: 149–158.
11. Gleba YY (1978) Microdroplet culture: tobacco plants from single mesophyll protoplasts. Naturwissenschaften 65: 158–159.
12. Gleba YY, Hoffmann F (1980) "*Arabiaobrassica*": A novel plant obtained by protoplast fusion. Planta 149: 112–117.
13. Jones LE, Hildebrandt AC, Riker AJ, Wu JH (1960) Growth of somatic tobacco cells in microculture. Am J Bot 47: 468–475.
14. Kao KN, Michayluk M (1975) Nutritional requirement for growth of *Vicia hajastana* cells and protoplasts at a very low population density in liquid media. Planta 126: 105–110.
15. Kao KN (1977) Chromosomal behaviour in somatic hybrids of soybean and *Nicotiana glauca*. Mol Gen Genet 150: 225–230.
16. Koop HU, Weber G. Schweiger HG (1983a) Individual culture of selected single cells and protoplasts of higher plants in microdroplets of defined media. Z Pflanzenphysiol 112: 21–34.
17. Koop HU, Dirk J, Wolff D, Schweiger HG (1983b) Somatic hybridization of two selected single cells. Cell Biol Int Rep 7: 1123–1128.
18. Koop HU, Schweiger HG (1985a) Regeneration of plants from individually cultivated protoplasts using an improved microculture system. J Plant Physiol 121: 245–257.
19. Koop HU, Schweiger HG (1985b) Regeneration of plants after electrofusion of selected pairs of protoplasts. Eur J Cell Biol 39: 46–49.
20. Koop HU, Spangenberg G (1989) Electric field-induced fusion and cell reconstitution with preselected single protoplasts and subprotoplasts of higher plants. In: Neumann E, Sowers A, Jordan C (eds) Electroporation and Electrofusion in Cell Biology, pp 355–366. New York, London: Plenum Press.
21. Kranz E, Bautor L, Lörz H (1990) In vitro fertilization of single, isolated gametes, transmission of cytoplasmic organelles and cell reconstitution of maize (*Zea mays L.*). In: Proceedings VII IAPTC Meeting, Amsterdam 1990, pp 252–257.
22. Krumbiegel G, Schieder O (1981) Comparison of somatic and sexual incompatibility between *Datura innoxia* and *Atropa belladonna*. Planta 53: 466–470.

23. Mejia A, Spangenberg G, Koop HU, Bopp M (1988) Microculture and electrofusion of defined protoplasts of the moss *Funaria hygrometrica*. Botanica Acta 101: 166–171.
24. Muir WH, Hildebrandt AC, Riker AJ (1954) Plant tissue cultures produced from single isolated cells. Science 119: 877–878.
25. Ogino T, Hiraoka, N, Tabata M (1978) Selection of high nicotine-producing cell lines of tobacco callus by single-cell cloning. Phytochemistry 17: 1907–1910.
26. Oksman-Caldentey K, Strauss A (1986) Somaclonal variation of scopolamine content in protoplast-derived cell culture clones of *Hyoscyamus muticus*. Planta Medica 1: 6–12.
27. Patnaik G, Cocking EC, Hasmill J, Pental D (1982) A simple procedure for the manual isolation and identification of plant heterokaryons. Plant Sci Lett 24: 105–110.
28. Potrykus I, Shillito RD (1986) Protoplasts: Isolation, culture, plant regeneration. In: Weissbach A, Weissbach H (eds). Methods in Enzymology, Plant Molecular Biology 118 pp 549–578. Orlando: Academic Press.
29. Potrykus I, Harms CT, Lörz H (1979) Multiple-drop-array (MDA) technique for the large-scale testing of culture media variations in hanging microdrop cultures of single cell systems. I. The technique. Plant Sci Lett 14: 231–235.
30. Reinert L (1956) Dissociation of cultures from *Picea glauca* into small tissue fragments and single cells. Science 123: 457–458.
31. Schäffler E, Koop HU (1990) Single cell nurse culture of tobacco protoplasts: physiological analysis of conditioning factors. J Plant Physiol 137: 95–101.
32. Schweiger HG, Dirk J, Koop HU, Kranz E, Neuhaus G, Spangenberg G, Wolff D (1987) Individual selection, culture and manipulation of higher plant cells. Theor Appl Genet 73: 769–783.
33. Spangenberg G, Koop HU, Schweiger HG (1985) Different types of protoplasts from *Brassica napus*: analysis of conditioning effects at the single cell level. Eur J Cell Biol 39: 41–45.
34. Spangenberg G (1986) Manipulation individueller Zellen der Nutzpflanze *Brassica napus* mit Hilfe von Elektrofusion, Zellrekonstruktion und Mikroinjektion. PhD thesis, Universität Heidelberg.
35. Spangenberg G, Schweiger HG (1986) Controlled electrofusion of different types of protoplasts including cell reconstitution in *Brassica napus*. Eur J Cell Biol 41: 51–56.
36. Spangenberg G. Koop HU, Schweiger HG (1986a) Microculture of single protoplasts of *Brassica napus*. Physiol Plant 66: 1–8.
37. Spangenberg G, Neuhaus G, Schweiger HG (1986b) Expression of foreign genes in a higher plant cell after electrofusion-mediated cell reconstitution of a microinjected karyoplast and a cytoplast. Eur J Cell Biol 42: 236–238.
38. Spangenberg G, Neuhaus G, Potrykus I (1990a) Micromanipulation of higher plant cells. In: Dix P (ed) Plant Cell Line Selection, pp 87–109. Weinheim: VCH Verlagsgesellschaft.
39. Spangenberg G, Osusky M, Oliveira MM, Freydl E, Nagel J, Pais MS, Potrykus I (1990b) Somatic hybridization by microfusion of defined protoplast pairs in *Nicotiana*: morphological, genetic, and molecular characterization. Theor Appl Genet 80: 577–587.
40. Spangenberg G, Freydl E, Osusky M, Nagel J, Potrykus I (1990c) Organelle transfer by microfusion of defined protoplast-cytoplast pairs. Theor Appl Genet (in press).
41. Stuart R, Street HE (1971) Studies on the growth in culture of plant cells. X. Further studies on the conditioning of culture media by suspensions of *Acer pseudoplatanus L*. J Exp Bot 22: 96–106.
42. Torrey JG (1957) Cell division in isolated single cells in vitro. Proc Natl Acad Sci USA 43: 887–891.
43. Vasil V, Hildebrandt AC (1965) Differentiation of tobacco plants from single, isolated cells in microcultures. Science 150: 889–892.
44. Weber G, Lark KG (1979) An efficient plating system for rapid isolation of mutants from plant cell suspension. Theor Appl Genet 55: 81–86.

Plant Tissue Culture Manual **B9**: 1–13, 1992.

Transformation and regeneration of maize protoplasts

CAROL A. RHODES & DAVID W. GRAY

Plant Biotechnology Department, Sandoz Crop Protection Corporation

Introduction

Several laboratories have successfully recovered mature plants from maize protoplasts [9, 11, 12, 15, 16]. In all cases, embryogenic suspension cultures were the source of competent protoplasts (those capable of division and plant regeneration). The limiting step has been to produce such suspensions repeatedly. Friable, embryogenic callus cultures were used as initial tissue for suspension cultures. The only commonality among the methods used to derive suspensions in these laboratories was that the donor suspensions seemed to require a minimum period (weeks to months) of growth in liquid culture before yielding competent protoplasts.

Genotype alone is not limiting, as success has been achieved with diverse genotypes, including a tropical inbred [11], a supersweet hybrid-derived haploid line [16], elite field corn lines [15] as well as a line derived from genetic mixtures [9]. This does not imply that any genotype will work as readily as any other, but rather that success is not dependent on one specific genotype.

Another difficulty is the frequency of partial or complete sterility in plants regenerated from protoplasts. This frequency seems to depend on the donor culture and may be related to the length of time *in vitro*. Frequency and vigor of plant regeneration directly from suspension cells often declines with increasing time in culture. These declines of regeneration vigor may be at least partly responsible for sterile plants regenerated from protoplasts. Recovery of plants in which there are problems with fertility may indicate genetic changes have occurred *in vitro* and/or regeneration protocols need to be improved for this trait.

Maize protoplasts appear capable of growing and dividing under a variety of culture conditions. Variations of both MS-based and N6-based media have been used successfully, in either liquid or solidified form. Addition of acetyl-salicylic acid improved plating efficiency in one case [2]. In several other cases, nurse cells have been shown to increase plating efficiency when used effectively [5, 7, 12, 16]. Choice of feeder cells is critical; among other desirable traits, cells must be growing rapidly to be effective.

Transformation of protoplasts can be accomplished by several means of direct DNA uptake. Two methods are used most often: electroporation or polyethylene glycol (PEG)-mediated uptake. Both methods have been used successfully with cereal protoplasts to produce stably transformed cells [1, 3,

6, 10]. Methods which combine both PEG and electroporation have been shown to increase transformation efficiency in some [14] but not all [18] systems. We describe here our procedure for electroporation [13].

Several genes have been used to select and identify stably transformed cells. These include genes conferring resistance to antibiotics, herbicides or other growth-inhibiting compounds. Examples of such genes are those encoding the enzymes neomycin phosphotransferase (NPT II), hygromycin phosphotrans-ferase (HPT), phosphinothricin acetyltransferase (PAT), and acetolactate synthase (ALS). Critical factors for any of these systems are timing of transfers, concentration of selective agent, and density of selected cells. Expression of the inserted gene must be sufficient to confer resistance to the selective agent. An example of one selection scheme using NPT II as the selectable marker gene is given here. Kanamycin has not always worked well in all cell systems; each case must be optimized and evaluated individually.

Procedures

Protoplast isolation

Steps in the procedure
1. Maize cell lines are maintained in liquid culture at 5 to 10 grams fresh weight per 50 ml of medium. Transfer cells twice weekly to fresh N6ap medium [12], N6 supplemented with 790 mg/l asparagine, 1.38 g/l proline, and 1 mg/l 2,4-dichlorophenoxyacetic acid (2,4-D).
2. Transfer suspension to a 50 ml graduated centrifuge tube (Corning #25330) and spin at 100 × g (Hermle Z320 centrifuge) for 5 minutes.
3. Pour off the supernatant. The packed volume of cells is approximately equal to grams fresh weight. Add 10 ml of Digestion Buffer for each gram of tissue. Generally, 5 grams of tissue is sufficient for a high yield of protoplasts (5 to 20×10^6).
4. Transfer the mixture to a 250 ml flask and place on a rotary shaker at 100 rpm for $1\frac{1}{2}$ to 2 hours in the dark.
5. After $1\frac{1}{2}$ hours, check a sample of the digest under a microscope for quality and approximate yield of protoplasts. Digest up to 1 hour longer if yield is below 1×10^6 per g.
6. Separate undigested cells from protoplasts by filtering the digest through a 149 μm polypropylene mesh screen (Spectrum Medial Industries #146432) and then through a 30 μm nylon mesh screen (SMI #146506). Rinse screens and collection beakers each time with 10 ml of Mannitol Buffer.
7. Distribute the protoplast suspension equally among four clear round-bottom centrifuge tubes (Nalge #3118, 50 ml polycarbonate). Add mannitol buffer to fill the tubes and centrifuge at 100 × g for 5 minutes.
8. Carefully pipette off the supernatant without disturbing the protoplast pellet. Discard the supernatant and add 15 ml of fresh Mannitol Buffer. Combine two tubes into one and centrifuge again at 100 × g for 5 minutes.
9. Repeat step 8 so that there is only one remaining tube with protoplasts. Spin and discard the supernatant. To prepare the protoplasts for electroporation, add 5 ml of Protoplast Culture Medium with the pH adjusted to 8.0. Estimate the number of viable protoplasts by staining a sample with 0.01% fluorescein diacetate [17].
10. Adjust the protoplast concentration to 3 to 4×10^6 per ml for transformation.

Direct uptake of DNA via electroporation

Steps in the procedure

1. Just prior to electroporation, protoplasts may be heat treated by submersing the tube in a 42 °C water bath for 3 minutes. This step should be followed as soon as possible by all subsequent steps to the dilution of protoplasts (step 5).
2. Transfer 0.5 ml aliquots of protoplasts into wells of a 24 well microtiter plate (Corning #25820). The number of wells loaded is determined by the number of protoplasts and the size of the experiment. A typical experiment is as follows:

Treatments	DNA (μg)	Voltage (V)	Capacitance (μF)
a) control	none	none	. . . .
b) control	none	225	1200
c) pZO3	30	200	1200
d) pZO3	30	225	1200
e) pZO3	30	250	1200

3. Add 75 μl of 2M KCl (final concentration 150 mM), 0.4 ml Protoplast Culture Medium pH 8.0, and plasmid DNA (10 to 50 μg per well) to each of the wells with protoplasts. Circular or linear DNA can be used.
4. Electroporate each well immediately after adding DNA and mixing with protoplasts (Hoeffer Scientific Instruments, Progenetor™). Optimum voltages may vary with protoplast size.
5. After treatments are electroporated, transfer protoplasts to 60 × 15 mm Petri plates and dilute with 2 volumes of Protoplast Culture Medium (final volume, 3 ml).

Note
4. Alternative electroporation apparatus: Bio-Rad Gene Pulser® with samples placed in cuvettes.

Growth of electroporated protoplasts

Steps in the procedure

1. Prepare feeder plates by suspending 0.5 g of BMS (Black Mexican Sweet suspension culture) cells in 1 ml of Protoplast Culture Medium per plate. Choice of feeder cells is important [7, 12].
2. Pipette 1.5 ml of feeder suspension mixture onto solidified Protoplast Culture Medium (100 × 15 mm Petri plates) and spread the mixture over the entire plate. Place a filter (Millipore # AABG 047 SO or HAEP 047 SW) over the feeder layer. Make 5 plates for each treatment.
3. Transfer 0.6 ml of protoplasts to each of the feeder plates and carefully pipette protoplasts over the surface of the filters.
4. Place finished plates in the dark or subdued light at 28 °C.

Selection of stable transformants

Steps in the procedure

1. One week after electroporation, transfer filters with protoplasts to plates with fresh culture medium at reduced osmolarity (N6ap with 0.1 M mannitol) containing 100 mg/l kanamycin. These plates may also contain feeder cells, made as before. A kanamycin-resistant feeder culture may be used, but is not essential.
2. Transfer filters after another week to N6ap supplemented with 1 mg/l 2,4-D and 100 mg/l kanamycin (no added mannitol) without any feeder cells.
3. Repeat weekly or biweekly transfers as needed, until rapidly growing colonies are distinctive among any background growth. This should take a total of 4 to 5 weeks after electroporation.
4. Callus can be assayed for presence of the introduced gene via enzyme assay [8] and DNA analysis.
5. Regenerate shoots by transferring callus to N6 or MS medium with reduced (0.05 to 0.25 mg/l) or no 2,4-D. Other hormonal and nutritional regimes which have been suggested include the use of cytokinins for a brief period [15] or increased sucrose concentrations [4].
6. Transfer shoots to clear jars or boxes (Magenta GA7) containing MS medium without any added hormones. It is important that shoots be grown under bright light all the way to maturity. Transfer to sterile soil mix after roots have grown out. Each genotype or cell culture may regenerate best under particular conditions which must be determined empirically.

Solutions
Mannitol Buffer

> 80 mM $CaCl_2$
> 0.5% w/v MES
> 0.30 M mannitol
> adjust pH to 6.0

Digestion Buffer

> To 100 ml of mannitol buffer, add:
> 2 g Cellulase RS (Yakult Pharmaceutical Co.)
> 100 mg Pectolyase Y23 (Seishin Pharmaceutical Co.)
> Filter sterilize. Stable at −20 °C.

N6 medium

100X Iron Stock

Na_2EDTA	3.72 g
$FeSO_4.7H_2O$	2.78 g

Boil the EDTA vigorously for 1-2 min in 200 ml H_2O. Add to the $FeSO_4$ dissolved in 200 ml H_2O and bring volume to 1 litre with H_2O.

10X N6 Major Salts

KNO_3	28.30 g
$MgSO_4$	0.90 g
$CaCl_2$	1.25 g
$(NH_4)_2SO_4$	4.63 g
KH_2PO_4	4.00 g

Add salts in order to a beaker with ca. 500 ml H_2O. Stir until dissolved and bring volume to 1 litre with H_2O.

100X N6 Minor Salts

$MnSO_4 \cdot H_2O$	0.33 g
H_3BO_3	0.16 g
$ZnSO_4 \cdot 7H_2O$	0.15 g
KI	0.08 g

Add salts in order to a beaker with ca. 500 ml H_2O. Stir until dissolved and bring volume to 1 litre with H_2O.

100X N6 Vitamin Stock

Thiamine-HCl	20 mg
Nicotinic acid	10 mg
Pyridoxine-HCl	10 mg
Glycine	40 mg
Casein hydrolysate	2 g

Dissolve in H_2O and bring to a final volume of 200 ml.

N6ap 1D Medium

Sucrose	30 g
L-asparagine	0.79 g
Proline	1.38 g
100X Iron stock	10 ml
10X N6 Major salts	100 ml
100X N6 Minor salts	10 ml
100X N6 Vitamins	10 ml
Inositol	2 ml of a 100 mg/ml stock
2,4-D	1 ml of a 1 mg/ml stock

Add sucrose, asparagine and proline to 800 ml H_2O and start mixing. Add remaining stock solutions. Bring volume to 1 litre. Adjust pH to 5.8 after all compounds are dissolved. For plates, add 7 g agar. Autoclave in volumes not greater than 1 litre for 20 min.

Composition of N6 medium

I. *Salts*   *Final Concentration*

Major elements	*mg/l*	*mM*
$(NH_4)_2SO_4$	463	3.5
KNO_3	2830	28.0
$CaCl_2$	125	1.13
$MgSO_4$	90	0.75
KH_2PO_4	400	2.94
Na_2 EDTA	37.2	0.20 (Na)
$FeSO_4\ 7H_2O$	27.8	0.10 (Fe)

Minor elements	*mg/l*	*μM*
H_3BO_3	1.6	25.8
$MnSO_4\ 1H_2O$	3.3	19.5
$ZnSO_4 7H_2O$	1.5	5.2
KI	0.8	5.0

Organic constituents	*mg/l*
inositol*	200
thiamine HCl	1.0
glycine	2.0
pyridoxine	0.5
nicotinic acid	0.5
cassein hydrolysate*	200
sucrose	30 g/l

Protoplast Culture Medium

N6ap with 2 mg/L 2,4-D
0.30 M mannitol
pH 5.8

* not in original formula as given by Chu *et al.*, 1975, Scient. Sinica 18:659.

References

1. Armstrong CL, Petersen WL, Buchholz WG, Bowen BJ, Sulc SL (1990) Factors affecting PEG-mediated stable transformation of maize protoplasts. Plant Cell Rep. 9: 335–339.
2. Carswell GK, Johnson CM, Shillito RD, Harms CT (1989) O-acetyl-salicylic acid promotes colony formation from protoplasts of an elite maize inbred. Plant Cell Rep. 8: 282–284.
3. Fromm ME, Taylor LP, Walbot V (1986) Stable transformation of maize after gene transfer by electroporation. Nature 319: 791–793.
4. Hodges TK, Kamo KK, Imbrie CW, Becwar MR (1986) Genotype specificity of somatic embryogenesis and regeneration in maize. Bio/Tech. 4: 219–223.
5. Kuang VK, Shamina ZB, Butenko RG (1983) Use of nurse tissue culture to obtain clones from cultured cells and protoplasts of corn. Fiziologiya Rastenii 30 : 803–812.
6. Lörz H, Baker B, Schell J (1985) Gene transfer to cereal cells mediated by protoplast transformation. Mol. Gen. Genet. 199: 178–182.
7. Lyznik LA, Kamo KK, Grimes HD, Ryan R, Chang K, Hodges TK (1989) Stable transformation of maize: the impact of feeder cells on protoplast growth and transformation efficiency. Plant Cell Rep. 8: 292–295.
8. McDonnell RE, Clark RD, Smith WA, Hinchee MA (1987) A simplified method for the detection of neomycin phosphotransferase II activity in transformed plant tissues. Plant Mol. Biol. Rep. 5: 380–386.
9. Morocz S, Donn G, Nemeth J, Dudits D (1990) Plant regeneration from haploid and diploid Zea mays (L.) protoplast cultures. International Plant Tissue Culture Congress, Amsterdam. Abstract A1–102.
10. Potrykus I, Saul M, Petruska J, Paszkowski J, Shillito R (1985) Direct gene transfer to cells of a graminaceous monocot. Mol. Gen. Genet. 199: 183–188.
11. Prioli LM, Sondahl MR (1989) Plant regeneration and recovery of fertile plants from protoplasts of maize (*Zea mays* L). Bio/Tech. 7: 589–594.
12. Rhodes CA, Lowe KS, Ruby KL (1988) Plant regeneration from protoplasts isolated from embryogenic maize cell cultures. Bio/Tech. 6: 56–60.
13. Rhodes CA, Pierce DA, Mettler IJ, Mascarenhas D, Detmer JJ (1988) Genetically transformed maize plants from protoplasts. Science 240: 204–207.
14. Shillito R, Saul M, Paszkowski J, Muller M, Potrykus I (1985) High efficiency direct gene transfer to plants. Bio/Tech. 3: 1099–1103.
15. Shillito RD, Carswell GK, Johnson CM, DiMaio JJ, Harms CT (1989) Regeneration of fertile plants from protoplasts of elite inbred maize. Bio/Tech. 7: 581–587.
16. Sun CS, Prioli LM, Sondahl MR (1989) Regeneration of haploid and dihaploid plants from protoplasts of supersweet (sh2sh2) corn. Plant Cell Rep. 8: 313–316.
17. Widholm JM (1972) The use of fluorescein diacetate and phenosafranine for determining viability of cultured plant cells. Stain Tech. 47: 189–194.
18. Yang H, Zhang HM, Davey MR, Mulligan BJ, Cocking EC (1988) Production of kanamycin resistant rice tissues following DNA uptake into protoplasts. Plant Cell Rep. 7: 421–425.

Plant Tissue Culture Manual C6: 1–12, 1992.

Virus elimination and testing

MARY C. COLEMAN[1] & WAYNE POWELL
Scottish Crop Research Institute, Invergowrie, Dundee, Scotland DD2 5DA
[1]*Agricultural Genetics Company Ltd., Cambridge CB4 4BH, UK*

Introduction

In recent years there has been an increasing awareness of the value of germplasm conservation. As new varieties comprise an increasingly large proportion of cultivated crops, the genetic diversity of major crop species is being eroded. A number of initiatives have been undertaken by international organisations to collect, evaluate and preserve wild and primitive genotypes. For seed-propagated crops this is relatively easy but many crops have to be propagated asexually because the plants do not produce seeds or the crop depends on the performance of a selected genotype. Thus asexually propagated material needs to be maintained clonally by an appropriate means of vegetative propagation.

Over the past 50 years *in vitro* multiplication methods have been developed to facilitate the rapid introduction of new varieties, international exchange of germplasm and the storage of material without the risk of infections which can occur *in vivo*. A prerequisite for the use of *in vitro* multiplication for germplasm storage and exchange is the availability of 'elite' disease-free plants from which cultures can be initiated. In this chapter will be discussed methods of detecting and eliminating virus infection from stock plants before or during micropropagation. It should also be borne in mind that such plants may be infected with bacteria and/or fungi and steps to eliminate them may also need to be considered.

Virus elimination

There are several methods available to produce virus-free plants, including heat treatment, meristem culture, heat treatment combined with meristem culture, adventitious regeneration in the presence of chemicals such as Virazole (1-D Ribofuranosyl-1,2,4 triazole-3-carboxamide) and culture of cells or protoplasts from non-infected cells.

Heat treatment is an effective way of inactivating isometric viruses [36], while meristem culture has gained importance due to the absence of virus in the apical dome in most species. The absence of virus has been attributed to competition in the meristem between production of virus particles and cell division, but it may also be due to a lack of vascular elements in the meristem, thus hindering the transport of the virus particles to the apical dome [36]. The absence of infected cells in mesophyll cell populations, and particularly in infections characterised by the presence of dark green areas within mosaic patterns, has offered the possibility of regenerating virus-free plants from such cells or areas [38, 15, 2, 34].

It is sometimes possible to suppress virus multiplication and eliminate viruses by the use of antiviral chemicals such as Virazole (syn Ribavirin) or vibarabine in the medium [43, 26, 29]. In a few cases such as lily [10] and apple [18] the use of Virazole has resulted in the production of virus-free plants. Incorporation of Virazole into potato explant and meristem culture media has been shown to give virus-free progeny from virus-infected explant and meristem donor plants [7]. Bittner *et al.* [4] showed that Ribavirin and 2,4-dioxohexa-hydro-1, 3, 5-triazine (DHT) significantly inhibited the replication of potato virus S. Meristem culture and adventitious regeneration in the presence of anti-viral chemicals is described below.

Methods of virus testing

Potential donor plants must be screened for the presence of infection, particularly if the virus is symptomless or if the plant is infected with more than one virus. The simplest methods involve the use of indicator plants which react promptly and characteristically to sap inoculation, usually with the formation of local lesions on the inoculated leaves [39]. The best 'general purpose' indicator plants are found in the Chenopodiaceae, particularly *Gomphrena globosa*. Virus particles can often be directly observed with an electron microscope, e.g. the quick leaf-dip method of Brandes [5]. The advantages of combining EM with serology have been elucidated by Gough & Shukla [17]. However, present methods limit the number of samples which can be examined. A widely used serological technique is enzyme linked immuno-sorbent assay (ELISA) [9]. It normally takes 1 to 2 days to complete and can handle very large numbers of samples.

More recent methods involving nucleic acid hybridisation to detect the viral genome are an alternative to ELISA. Nucleic acid hybridization has the advantage of greater sensitivity and a high degree of specificity.

Symptoms

Symptoms of viral infection vary from mosaics or ringspots to variegation, vein-banding or clearing. Internal symptoms may also be present, i.e. intra-cellular inclusions which are aggregates of virus particles large enough to be seen using an optical microscope. The two main types of inclusion are crystalline and amorphous, the latter being known as X bodies. Other structures may occur in infected cells, such as 'pinwheels', which are symptoms of infection with potato virus Y, although it is unlikely that they consist of virus

[19]. The symptoms associated with virus infections have been reviewed by Smith [39], Esau [14] and Matthews [30].

Test plants

Test plants are of considerable importance not only for detecting and identifying known viruses but also for revealing the presence of new ones. They are also of use in differentiating viruses and in separating certain virus complexes such as those that commonly occur in potato [20, 21, 11]. Viruses occurring in a complex can be separated by observing their reaction when challenged by a test plant. Test plants may be primary, secondary or tertiary separators, depending on whether in the course of a stepwise virus separation they are suitable for the simultaneous separation of two viruses (primary separator) or for the further separation of residual virus complexes (secondary and tertiary separators). A check-list of host plants for identifying and separating twelve potato viruses is given by Horvath [22].

Virus detection may be seasonally dependent, e.g. in *Pelargonium* it is most efficient in early spring, using mechanical inoculation of sap diluted in a phosphate buffer on *Chenopodium quinoa*. This is sensitive to all known *Pelargonium* viruses [44].

Electron microscopy

Viruses can be observed in crude sap using electron microscopy. The sample is mounted on a 3 mm copper grid covered with a thin film of plastic (Formvar). The grids contain a number of apertures (60 to 160 meshes/cm). The virus particles stick to the Formvar film, but as they are highly transparent to electrons shadow-casting or negative staining is essential [35]. Shadow casting may be performed with dried preparations, which can be prepared elsewhere and sent to a suitably equipped laboratory for examination.

Serology

The antigenic properties of viruses represent the single most useful criterion for their reliable detection and identification. Early serological detection techniques included the micro-precipitation test, the chloroplast agglutination test and the Ouchterlony agar double-diffusion test [35]. Enzyme-linked immunosorbent assay (ELISA) is a sensitive test which is widely used. Most workers use the direct double-antibody sandwich method which requires the preparation of a different antibody conjugate for each virus to be tested [8]. It is the method of choice when it is important to distinguish between closely related serotypes. It requires the preparation of one antiserum derived from a single animal species. The indirect ELISA method requires the use of two antisera prepared in different animal species (e.g. rabbit and chicken) and a single enzyme conjugate can be used for all virus systems. The indirect ELISA methods can detect a broader range of serologically related viruses and are usually more sensitive than the direct methods [40]. Multi-layered sandwich procedures are often the most sensitive, e.g. using an antibody from chicken, a viral antigen, an antibody from mouse, rabbit anti-mouse globulins and enzyme-labelled goat anti-rabbit globulins. The use of an avian antibody in one of the layers is advantageous [1] as the absence of cross-reaction between avian and mammalian globulins keeps background readings low [42].

Observing serological reaction with an electron microscope provides another means of virus identification. Derrick [12] applied the principle of a solid phase immunoassay to electron microscopy by trapping viruses on grids coated with specific antiserum. This technique is known as ISEM [37], and has been found to have a sensitivity similiar to that of ELISA. By pre-treating the grids with protein A (PALIEM), the antibody molecules can be anchored more efficiently and the sensitivity of virus detection may be slightly improved [28]. Serological procedures for detection of plant virus infections are described and reviewed by Van Regenmortel [41].

Presence of double-stranded RNA

The use of viral double-stranded (ds) RNA for disease detection arises from the fact that only plants infected with RNA viruses or virus-like agents contain homogeneous segments of high molecular weight ($70.1 + 10^6$ KDa) ds RNA [23, 31, 32]. Additional diagnostic information about the virus can be obtained from the pattern of genomic and subgenomic ds RNA species upon gel electrophoretic analysis of the isolated ds RNAs. For procedures see Morris & Dodds [31], Dodds & Bar-Joseph [13], Jordan *et al.* [25], Jordan & Dodds [24].

Nucleic acid probes

Nucleic acid hybridisation methods have been used to detect viral genomes with the cDNA probes capable of identifying viral target sequences in the 1 to 10 ng range. The method involves the hybridisation of ^{32}[P] labelled cDNA clones with crude sap spots which have been immobilised on a nitrocellulose membrane, and has been described by Baulcombe *et al.* [3] for potato virus X. The cDNA is prepared by reverse transcription of partially purified viral RNA, and the probes are radioactively labelled with ^{32}[P] by nick translation. Non-radioactively labelled probes such as biotin-labelled nucleotides and the fluorescein-conjugated avidin system overcome the safety problems associated with the use of radioactivity [27].

Conclusions

Detection and elimination methods should be rapid, reliable and sensitive and should at the same time give clear-cut results. Schemes for the production of virus-free planting stock and subsequent multiplication and distribution have been operating for some vegetatively-propagated horticultural crops in agriculturally advanced countries for many years. For example, virus-indexed strawberry plants have been available for over 50 years in Britain where the Nuclear Stock Association Ltd distributes virus tested material of flower bulbs, soft fruit and fruit trees [6]. The methods of detection depend on both the plant and the virus, and should use a combination of simple biological tests and more advanced tests such as ELISA and cDNA probes. Meristem culture and heat treatments should result in viral elimination.

The following sections outline some of the methods which can be used in virus elimination and testing.

Protocols

Meristem culture

1. Select starting material from seedlings, new buds or young shoots.
2. Excise apical and lateral buds.
3. Surface sterilise (optional) with 80% v/v alcohol for 30 seconds followed by 1% v/v sodium hypochlorite for 15 minutes. Rinse in sterile distilled water 3 times. All procedures after the alcohol sterilization are carried out in a laminar flow cabinet.
4. With the aid of a binocular microscope and using sterile forceps and scalpel remove the outer leaves and primordia until the glossy apical dome is visible.
5. Excise the dome and place on the culture medium.
6. Seal the tube/jar and place in a growth room with 16 h photoperiod, 160–180 μmol photons $m^{-2} s^{-1}$.

Notes

2. To eliminate viruses the size of the dome should not exceed 0.2–0.5 mm in length, depending on the plant species.
3. Surface sterilisation is optional and is not necessary with most material as the apical dome is normally very well protected and free from surface contaminants.
5. The composition of the nutrient medium is important. Each plant species and sometimes different cultivars within a species require special media. Murashige & Skoog [33] medium and Gamborg's B5 [16] are the most widely used.
   Some excised meristems (e.g. *Cattleya*) produce toxic substances, and in these cases a liquid medium should be used and the meristem placed on paper bridges.
   The percentage of isolated meristems developing into virus free plants is generally small.
   By culturing meristems in the presence of a suitable antiviral chemical it is possible to maintain inhibitory conditions for long enough to eliminate the virus.
6. The combination of heat treatment of the plant with meristem culture is more efficient if viruses are heat sensitive. Heat treatment varies; plants can be exposed to temperatures of 32–35 °C for 23 days, to 33–37 °C for 4 weeks or to 37–38 °C for 20 to 40 days depending on the plant and virus involved. The temperature and treatment time should be chosen to allow the plant to just survive while inactivating the virus. Heat treatment and meristem culture are commonly used with potato (*Solanum tuberosum*), carnation, strawberry and chrysanthemum.

Virus elimination using differentiating systems

1. Take internodal explants 2 cm in length from an actively growing donor plant.
2. Surface sterilise in 80% v/v alcohol for 30 sec followed by 1% v/v sodium hypochlorite for 15 minutes. Rinse three times in sterile water.
3. Cut off approximately 3 mm from each end of the explant and divide it longitudinally.
4. Place cut surface down on suitable regeneration medium containing an antiviral chemical e.g. 250 μm Ribavirin.

4. Ribavirin must be continuously present in the medium for virus elimination to occur.
   Plants take longer to differentiate and grow in the presence of Ribavirin.

ELISA [9]

1. Put 200 µl of antibody ($\propto$ globulin) 1 µg/ml in coating buffer into each cell of a 96 well microtitre plate. Incubate for 2 hours at 37 °C.
2. Prepare samples by grinding in extraction buffer. Wash plate with phosphate buffered saline (PBS) plus Tween by flooding the plate and leave it for 3 minutes. Repeat the process three times.
3. Add 200 µl of sample in duplicate to the plate and incubate overnight at 4 °C.
4. Wash the plate 3 times.
5. Add 200 µl antibody-conjugate (enzyme labelled $\propto$ globulin) and incubate at 37 °C for 3—6 hours.
6. Wash plate 3 times as in (2) above.
7. Add 200 µl freshly prepared substrate to each well. Incubate at room temperature (20—24 °C) for 1 hour.
8. Stop reaction with 50 µl of 3 M solution of NaOH.
9. Assess results (a) visual observation of colour reaction
   (b) measure absorbance at 405 nm.

Notes
1. Many commercial ELISA kits are available to detect viruses in agricultural and horticultural crops e.g. those by Boehringer (Mannheim).
   Preparation of antiserum is described by Noordan [35] and Clark & Adams [9].

Solutions
- Coating buffer:
  - NaHCO$_3$ 2.93 g ⎫ in 1 litre, pH 9.6
    Na$_2$CO$_3$ 1.59 g ⎬
- Phosphate buffered saline (PBS):
  - Na$_2$HPO$_4$.2H$_2$O (0.01 M)   8.9 g ⎫
  - NaH$_2$PO$_4$.2H$_2$O (0.01 M)   7.8 g ⎬ in 1 l, pH 7.0
  - NaCl   42.5 g, ⎭
- Extraction buffer:
  - PBS – Tween
  - 2% w/v polyvinylpyrrolidone (PVP)
- Washing buffer:
  - PBS – Tween (= PBS + 0.5 ml/l Tween)
- Conjugate buffer:
  - PBS – Tween
  - 2% w/v PVP
  - 0.2% w/v Ovalbumin
- Substrate buffer:
  - 0.1 M diethanolamine, pH 9.8

- Substrate:
  - 0.6 mg/ml 4-nitrophenylphosphate in substrate buffer
  Siliconise all glassware used for antiserum.

Protein A linked immuno-electron microscopy (PALIEM)
1. Coat copper electron microscope grids with 0.4% w/v Formvar in water-free ethylene dichloride.
2. Grip grid, coated side uppermost, with forceps and place a 5 μl drop of protein A (0.1 mg/l in 5 mM sodium phosphate buffer, pH 7.0) on it.
3. Incubate for 10 minutes at 20 °C.
4. Wash grid with approximately 20 drops of sodium phosphate buffer (5 mM, pH 7.0).
5. Place 5 μl of phosphate buffer on the grid and dip the cut edge of a leaf into it (incubate as above).
6. Wash grid with buffer (4).
7. Add 5 μl of dilute antiserum and incubate (3).
8. Wash the grid with buffer followed by distilled water (30 drops of each).
9. Stain by placing a drop of 2% w/v uranyl acetate in water on the slide.
10. Examine using an electron microscope for virus presence.

References

1. Al Moudallal Z, Altschuh D, Briand JP & Van Regenmortel MHV (1984) Comparative sensitivity of different ELISA procedures for detecting monoclonal antibodies. Journal of Immunological Methods 68: 35–43.
2. Atkinson PH & Matthews REF (1970) On the origin of dark green tissue in tobacco leaves infected with tobacco mosaic virus. Virology 90: 344–356.
3. Baulcombe D, Flavell RB, Boulton RE & Jellis GJ (1984). The sensitivity and specificity of a rapid nucleic acid hybridisation method for the detection of potato virus X in crude sap samples. Plant Pathology 33: 361–370.
4. Bittner H, Schenk G, Schuster G & Kluge S (1989) Elimination by chemotherapy of potato virus S from potato plants grown *in vitro*. Potato Research 32: 175–179.
5. Brandes J (1957) Eine elektronenmikroskopische Schnellmethode zum Nachweis faden und Stabchenformiger, Viren, insbesondere in Kartoffeldunkelkeimen. Nachr Bt Dr Pflschutzd 9: 157–152.
6. Brunt AA (1985) The production and distribution of virus-tested ornamental bulb crops in England: Principles practice and prognosis. Acta Horticulturae 164: 153–161.
7. Cassells AC & Long RD (1982) The elimination of potato viruses X, Y, S and M in meristem and explant cultures of potato in the presence of Virazole. Potato Research 25: 165–173.
8. Clark MF (1981) Immunosorbent assays in plant pathology. Annual Review of Phytopathology 19: 83–106.
9. Clark MF & Adams AN (1977) Characteristics of the microplate method of enzyme linked immunosorbent assay for the detection of plant viruses. Journal of General Virology 34: 475–483.
10. Cohen AJ (1986) Plant tissue cell culture abstracts. International Congress 6: 30.
11. De Bokx JA (1972) Viruses of potato and seed-potato production. Wageningen: Centre for Agricultural Publishing and Documentation.
12. Derrick KS (1973) Quantitative assay for plant viruses using serologically specific electron microscopy. Virology 56: 652–653.
13. Dodds JA & Bar-Joseph M (1983) Double-stranded RNA from plants infected with clostero viruses. Phytopathology 73: 419–423.
14. Esau K (1967) Anatomy of plant virus infections. Annual Review of Phytopathology 5: 45–76.
15. Fulton, R.W. (1951) Superinfection by strains of tobacco mosaic virus. Phytopathology 41, 579–592.
16. Gamborg DL, Miller RA & Ojima K (1968) Nutrient requirements of suspension cultures of soybean root cells. Experimental Cell Research 50: 151–158.
17. Gough KA & Shukla DD (1980) Further studies on the use of protein A in immuno electron microscopy for detecting virus particles. Journal of General Virology 51: 415–419.
18. Hansen AJ & Lane WD (1985) Elimination of apple chlorotic leafspot virus from apple shoot cultures. Plant Diseases 69: 134–135.
19. Hiebert E, Purcifull DE, Christie RG & Christie SR (1971) Partial purification of inclusions induced by tobacco etch virus and potato virus Y. Virology 43: 638–646.
20. Horvath J (1963) Neuere Beitrage zum Vorkommen von Kartoffelviren mit besonderer Rücksicht auf Komplexinfektionen. Acta Agronomica Academicae Scientiarum Hungarice 14: 67–81.
21. Horvath J (1967) Separation and determination of viruses pathogenic to potato virus Y. Acta Phytopathologica Academiae Scientiarum Hungaricae 2: 319–360.
22. Horvath J (1985) A check-list of new host plants for identification and separation of twelve potato viruses. Potato Research 28: 71–89.
23. Ikegami M & Fraenkel-Comrat H (1979) Characterisation of double stranded ribonucleic acid in tobacco leaves. Proceedings National Academy of Sciences USA 76: 3637–3640.
24. Jordan R & Dodds JA (1985) Double-stranded RNA in detection of diseases of known and unproven viral etiology. Acta Horticulture 164: 101–108.

25. Jordan RL, Heick JA, Dodds JA & Ohr H (1983) Rapid detection of sunblotch viroid RNA and virus-like double-stranded RNA in multiple avocado samples. Phytopathology 73: 791.(Abs.)
26. Kartha KK (1986) Production and indexing of disease-free plants. In: Plant tissue culture and its agricultural application. (LA Withers and PG Alderson, Eds). London, Butterworth Press.
27. Lange L (1986) The practical application of new developments in test procedures for the detection of viruses in seed. In: Developments in Applied Biology 1. Developments and Applications in Virus Testing. Eds RAC Jones and L Torrance. Association of Applied Biologists, Wellesbourne, UK.
28. Lesemann DE & Paul HL (1980) Conditions for the use of protein A in combination with the Derrick method of immuno electron microscopy. Acta Horticulturae 110: 119–128.
29. Long RD & Cassells AC (1986) Elimination of viruses from tissue culture in the presence of antiviral chemicals. In: Plant Tissue Culture and its Agricultural Application (LA Withers and PG Alderson, Eds), Butterworth Press.
30. Matthews REF (1970) Plant Virology, Academic Press, New York.
31. Morris TJ & Dodds JA (1979) Isolation and analysis of double-stranded RNA from virus infected plant and fungal tissue. Phytopathology 69: 854–858.
32. Morris TJ, Dodds JA, Hillman B, Jordan RL, Lommel SA & Tamalki SJ (1983) Viral specific ds RNA: diagnostic value for plant virus disease identification. Plant Molecular Biology Reporter 1: 27–30.
33. Murashige T & Skoog F (1962) A revised medium for rapid growth and bioassay with tobacco tissue cultures. Physiological Plantarum 15: 473–497.
34. Muraskishi HH & Carlson PS (1976) Regeneration of virus-free plants from dark green islands of tobacco mosaic virus-infected tobacco leaves. Phytopathology 66: 931–932.
35. Noordam D (1973) Identification of plant viruses. Methods and experiments. Centre for Agricultural Publishing Documentation, Wageningen.
36. Quak, F. (1977). Meristem culture and virus-free plants. In: Applied and fundamental aspects of plant cell, tissue and organ culture. J Reinert and YPS Bajaj (Eds). Springer-Verlag, Berlin.
37. Roberts IM, Milne RG, Van Regenmortel MHV (1982) Suggested terminology for virus-antibody interactions observed by electron microscopy. Intervirology 18: 147–149.
38. Shepard JF (1975) Regeneration of plants from protoplasts of potato virus X infected tobacco leaves. Virology 66: 492–501.
39. Smith KM (1974) Plant Viruses. Great Britain Northumberland Press.
40. Van Regenmortel MHV (1982) Serology and immunochemistry of plant viruses. Academic Press, New York.
41. Van Regenmortel MHV (1985) New serological procedures including the development and uses of monoclonal antibodies in virus detection and diagnosis. Acta Horticulturae 164: 187–194.
42. Van Regenmortel MHV (1986) The potential for using monoclonal antibodies in the detection of plant viruses. In: Developments in Applied Biology 1. Developments and Applications in Virus Testing. Eds RAC Jones and L Torrance. Association of Applied Biologists, Wellesbourne, UK.
43. Walkey DGA (1980) Production of virus-free plants by tissue culture. In: Tissue culture methods for plant pathologists (DS Ingram and JP Helgeson, Eds), 109–117. Blackwell Scientific Publications, Oxford.
44. Welvaert W & Samyn G (1985) Relative importance of Pelargonium viruses in cutting nurseries. Acta Horticulturae 164: 195–198.

Plant Tissue Culture Manual **D7**: 1–20, 1992.
© 1992 *Kluwer Academic Publishers.*

Isolation and uptake of plant nuclei

PRAVEEN K. SAXENA[1] & JOHN KING[2]

[1] *Department of Horticultural Science, University of Guelph, Guelph, Ontario, Canada, N1G 2W1;*
[2] *Department of Biology, University of Saskatchewan, Saskatoon, Saskatchewan, Canada, S7N 0W0*

Introduction

Genetic transformation of animal cells by transplantation of isolated organelles, nuclei, and chromosomes is well documented and has played a significant role in investigating chromosome mapping and the regulation of gene expression [11, 13]. In plants, direct DNA transfer using isolated organelles was attempted following the discovery of techniques to eliminate the plant-specific barrier, the cell wall, which previously hampered the introduction of organelles into cells. Predictably, as soon as plant cells succumbed to procedures capable of enzymic degradation of cell walls, a whole new concept of handling naked plant cells (protoplasts) emerged as well as a genuine hope of being able to transform plant cells by foreign DNA introduction [15]. Thus, many workers studied the uptake of a variety of macromolecules and demonstrated the ability of plant protoplasts to accept foreign particles such as ferritin, bacteria, and isolated organelles like nuclei and chloroplasts (see references 7, 14, 21 for extensive reviews).

The selective transfer of large amounts of DNA into the protoplasts via a whole nucleus, chromosome, or a part thereof as opposed to the fusion of two entire protoplasts holds great potential in developing ways to improve commercial crops. For example, many traits of economic significance such as yield and tolerance to salinity, drought, and extreme temperatures, are encoded by more than one gene. Although impressive progress has been made recently in transferring foreign genes into plant cells, a number of problems associated with the application of this technology for crop improvement still remain to be resolved, particularly the number of genes which can be transferred.

The current gene transfer strategies only allow the transfer of one or two genes using either indirect (*Agrobacterium*-mediated) or direct (plasmid DNA) methods of transformation. Construction of plasmids harbouring multiple genes at present is difficult and it does not seem likely that such large plasmids could be transferred effectively. Until such time that the genes regulating characters of economic importance are identified, cloned, and vehicle-bound (in plasmids), the transfer of isolated organelles with mapped and identified genes of interest would seem to be an attractive approach in incorporating polygenic traits. Furthermore, nuclear transplantation allows selective transfer of nuclei in an unmodified cytoplasm, that of the recipient, as opposed to protoplast fusion which results in mixing of cytoplasms as well. Selective transfer of nuclei may help avoid the problems of cytoplasmic incompatibility and the transfer of undesirable traits of cytoplasmic origin. Another important application of organelle transfer could be the induction of genetic variability into highly inbred germplasms particularly in self-pollinated crops.

In this chapter, the techniques for the isolation of nuclei and their transfer into plant protoplasts will be discussed.

Isolation of nuclei

For an efficient nuclear transplantation experiment, the nuclei should be available in large numbers. It is also important that nuclear preparations be free of cytoplasmic contamination and contain morphologically intact and biologically active nuclei. These fundamental requirements have been the issue of several investigations over the years [3, 22, 23, 28, 29]. There seems to be a general agreement that source material should be soft and having a minimum of starch, phenols, and tannins. Various favourable sources recognized include meristematic, etiolated, or embryonic cells. In this regard, cell suspension cultures seem particularly suitable as they provide large populations of actively growing cells which are an excellent source of protoplasts. Protoplasts have been found to be the most suitable source for nuclei isolation because of the absence of cell walls which makes it easier to release the nucleus without contamination by cell wall fragments.

Protocol 2 (see later) describes the method developed in our laboratory for isolating nuclei from protoplasts of cell suspensions. To achieve good nuclear preparations using protoplasts, a controlled denaturing of cell membrane is essential. In this connection, Triton X-100 is the common detergent used to solubilize cell membranes and to remove other cytoplasmic debris. However, the concentrations found to be effective (0.1–1%) in rupturing the protoplasts were associated with deleterious effects on the nuclear envelope [9, 10]. A solution to this problem has been presented [22, 23] by employing a two-step destabilization of protoplast membranes. Firstly, the protoplasts are deplasmolysed in a hypotonic nuclei isolation buffer containing a low Triton X-100 concentration (0.01–0.02%). In the second step, these deplasmolysed protoplasts are homogenized in a glass homogenizer by applying 10–15 gentle strokes or by passing through needles of 18–26 gauge. With this modification in the procedure, Saxena et al. [22] were able to obtain clean nuclei (Fig. 1A) which retained a high degree of nuclear envelope integrity. The most crucial factor which allowed the lysis of protoplasts using lower concentrations of Triton X-100 was the pH of the isolation buffer. A narrow range of pH, between 5.2 and 5.5 only, was found to be suitable. At pH values lower or higher than 5.2–5.5, the nuclear yields as well as the quality of the nuclei were unsatisfactory.

Once isolated, nuclei need to be stabilized osmotically in order to insure against lysis and membrane damage. Various stabilizers that have been used include sucrose, mannitol, dextran, and ficoll. Divalent cations like Mg^{2+} and Ca^{2+} have also been used as protectants but they cause clumping of nuclei, but this can be avoided by replacing these with Na^+ and K^+ [22, 29]. Other compounds which have been used for the overall good nuclear preparation are ethylene diamine tetracetic acid (EDTA) to block the action of DNAse in the preparation [1], and to facilitate protoplast lysis [12], and polyvinylpoly-pyrrolidone, diethyldithiocarbamate, and mercaptobenzthiazol when added to the isolation buffer help remove phenolics and inhibit phenol oxidases.

Some additional thoughts on nuclei isolation

1. While preparing protoplasts for subsequent nuclei isolation, optimization of the type and concentration of enzymes used for cell wall digestion is important as the requirements vary from one system to another. It is also worth looking into the relationship between conditions of protoplast isolation and rupture. In the first step, the emphasis is given on creating conditions for maximal yield and stability of protoplasts but the opposite is the aim in the next step where the nuclei are released by inducing rupture of protoplasts. It may be useful to isolate protoplasts in isolation solutions which favour high yields but relatively low stability of protoplasts. The stability of protoplasts refers to their ability to remain intact and to survive repeated centrifugations necessary for their purification which may result in their lysis. In certain cultures (e.g. *Vicia hajastana*), protoplasts isolated in the presence of mannitol as an osmoticum are relatively more stable than those isolated with sucrose or a mixture of NaCl and KCl. In our recent experiments (unpublished), when mannitol was replaced by sucrose or salts as the osmoticum during isolation, subsequent rupture of protoplasts produced very little debris providing a much cleaner nuclear preparation. A variation in the type of cell wall digesting enzymes, the pH of the isolation mixture, and the incubation temperature may be rewarding as these factors greatly influence protoplast stability.
2. In order to obtain nuclei with maximum nuclear envelope integrity, the detergent concentration should be optimized. Optimization in this case refers to finding the minimal effective concentration of the detergent and the optimum pH which allows the effective rupture of the protoplasts. An increase in the period of deplasmolysis from 5 to 15 min in the NIB (Protocol 2) may prove beneficial in situations where protoplasts are tough to break at lower detergent concentrations.

Uptake of Nuclei

Experiments to induce nuclear uptake as a method of gene transfer were first conducted by Potrykus and Hoffman [20] who were able to introduce foreign nuclei in about 0.5% of the protoplasts by using membrane modifiers (lysozyme, sodium nitrate) together with centrifugation. The percentage of uptake of nuclei by protoplasts was considerably improved (to about 5%) when the uptake was induced using PEG, high pH and Ca^{2+} [16]. The determination of uptake frequencies in the above studies was based on microscopic estimations of host protoplasts containing the transferred nuclei stained with a fluorescent dye. For an accurate estimation of uptake frequencies, it is essential to be able to differentiate the nuclei actually taken-up by protoplasts from those merely attached to the surface of the protoplasts. In previous studies, the location of the nuclei, i.e. outside or inside the protoplasts, was determined by rolling a sample of the protoplast population after uptake between the slide and the coverslip, a process which may not always be accurate. Saxena *et al.* [25] devised a simple procedure based on differential staining of nuclei to follow the uptake of nuclei (Fig. 1B). In this method, nuclear uptake is carried out using the nuclei stained with a fluorescent dye (Hoechst 33258). The entire population of protoplasts is then stained with Evan's Blue. The nuclei attached to the surface of the protoplasts appear blue when examined under a light microscope in bright field and the nuclei taken-up by protoplasts show fluorescence under UV light (Fig. 2). Uptake frequencies ranging from 4 to 6% were obtained with this procedure [25].

The mechanism of nuclear uptake by protoplasts is not well understood, but various possibilities can be discussed based upon indirect evidence available from other studies. Ferritin, representing the smaller class of particles (0.1 μm) has been shown to gain access to protoplasts through endocytosis via coated vesicles. Particles in the middle range (0.1–3 μm), such as polystyrene latex spheres and bacteria, were accepted by membrane invagination and endocytosis. Larger particles (3–16 μm), such as protoplasts, have been shown to be taken up by fusion (for a review see 6). Therefore, nuclear uptake seems likely to occur either by fusion or through plasma membrane invagination by endocytosis (Fig. 3). The size of the nuclei to be introduced relative to the size of recipient protoplast may determine the mode of uptake which may involve invagination of plasma membrane in the case of smaller nuclei and fusion where the donor and recipient are not very different in size. The events following the uptake of nuclei into protoplasts are also unknown. Figure 3 shows various possible ways in which transfer and integration of nuclear DNA may occur.

Integration of nuclear DNA

In the first investigation to follow the integration of the foreign nuclei introduced into protoplasts, Lörz and Potrykus [16] utilized two chlorophyll-deficient, light sensitive tobacco mutants referred to as *sublethal* and *virescent*, as donors of nuclei and recipient protoplasts. These recessive mutants were previously shown to complement successfully by protoplast fusion [17]. The hybrids recovered following transplantation of *virescent* nuclei into *sublethal* protoplasts were identifiable from the parental cell colonies under high light, but the plants regenerated from such cell colonies were all of a *sublethal* type, hence no biological proof could be gathered regarding the integration of the transferred nuclear genes [16]. Success in this area has since been reported by Saxena *et al.* [24] who induced nuclear uptake into protoplasts using PEG treatment and gathered convincing evidence to show that following nuclear uptake, genetic integration did occur in the recipient cells. They [24] utilized, as the recipient, a *Datura innoxia* auxotrophic mutant (*Pn-1*) which had an absolute requirement for pantothenate for growth, and *Vicia hajastana* as the donor of nuclei. Following uptake of *Vicia* nuclei, prototrophic clones were selected on a medium lacking pantothenate. When subjected to dot blot hybridization, the genomic DNA of the putative transformed prototrophic clones showed the presence of donor (*Vicia*) DNA providing evidence that *Vicia* DNA did integrate into the host genome following the uptake. Further, the restoration of morphogenic ability in a non-morphogenic cell line also strengthened the confidence that genetic transfer and integration only could account for this observation. It should, however, be noted that in this study, the correction of auxotrophy was used as the selective marker, a particularly stringent selection pressure. It is noteworthy that in all previous successful attempts on nuclear transplantation, in yeast and mammalian cells, auxotrophy was the marker used in selection [2, 4, 13].

Perhaps the key to successful nuclear transplantation is the choice of an appropriate marker for selection. Unfortunately, a transformation system describing the transfer of specific gene(s) via nuclei as carriers has not been developed yet. The use of other selective markers commonly employed in model plant gene transfer systems, such as the resistance to an antibiotic (e.g. kanamycin), may not be very effective in nuclear transplantation experiments because introduced nuclei would release large amounts of undefined, but very little of specific, DNA. For example, the nuclei isolated from cells of transgenic kanamycin-resistant plants are likely to contain, at best, a few copies of the gene which confers the resistance. Such a low copy number, coupled with generally low frequency of nuclear uptake (compared to up to 90% with other methods), and a reduced viability of the host protoplasts after uptake treatment, pose serious problems in recovering nuclear hybrids. In this context, it may be rewarding to use markers such as kanamycin, chlorsulfuron, and methotrexate, if nuclei are isolated from cells containing multiple copies of the marker gene [30].

In addition to the choice of marker, the relative sizes of the recipient protoplasts and transferred nuclei may also be critical in the recovery of transformed cells. Larger protoplasts are more likely to survive repeated centrifugations required during purification steps prior to culturing when they contain one or more introduced nuclei. The presence of starch grains in the cytoplasm, the density of the cytoplasm, and the number of other organelles in the protoplast (particularly the chloroplasts) will also affect the survival of recipients through various cultural manipulations requiring centrifugation. A high frequency of survival and further development of recipient protoplasts is essential for successful isolation of transformed clones considering the low frequency of nuclear uptake. Thus, the protoplasts from cell suspension cultures which have less dense cytoplasm and no or fewer chloroplasts are ideal recipients.

Another facet of nuclear-based, limited but discrete, gene transfer is the formation of micronuclei, which are essentially spheres containing only a few to single chromosomes surrounded by a thin layer of cytoplasm and plasma membrane. In experiments with animal cells, the fusion of these structures with the desired recipient results in hybrid clones containing only one or a few introduced chromosomes [5]. The induction of micronuclei and their isolation from plant cells has been reported [26]. However, the transfer and integration of foreign DNA by transplanting micronuclei into host cells or protoplasts has not been achieved. On the other hand, the efforts to isolate and transplant chromosomes into plant cells have met with some success. Many workers have reported the successful isolation of chromosomes from a variety of cell cultures e.g. *Triticum, Papaver, Vicia, Petunia,* [9, 18]. The uptake of wheat and parsley chromosomes by wheat, parsley, and maize protoplasts has been described with convincing cytological evidence of their incorporation into the recipient protoplasts [27]. Recently, Griesbach [8] reported the transfer of isolated *Petunia alpicola* chromosomes to *P. hybrida* protoplasts by microinjection. The transformants were shown to have the donor-specific marker proteins and the enzymes of flavonoid pathway as evidence of gene transfer. The characteristic flavonoid enzymes incorporated in the recipient species were transmitted in Mendelian fashion [8].

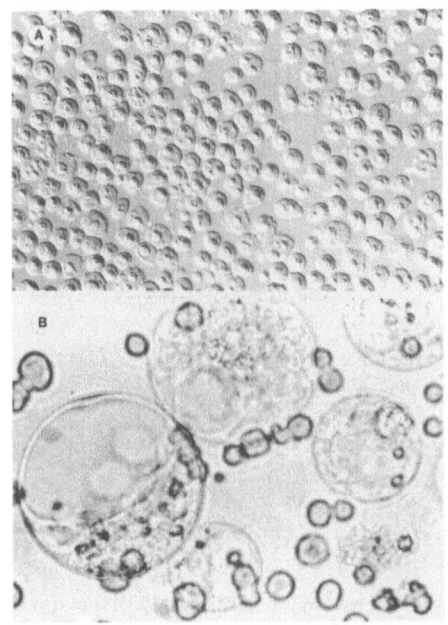

Fig. 1. Isolation and uptake of nuclei. A. Nuclei isolated by the procedure described in protocol 2. B. Induction of nuclear uptake by protoplasts in the presence of uptake inducing solution.

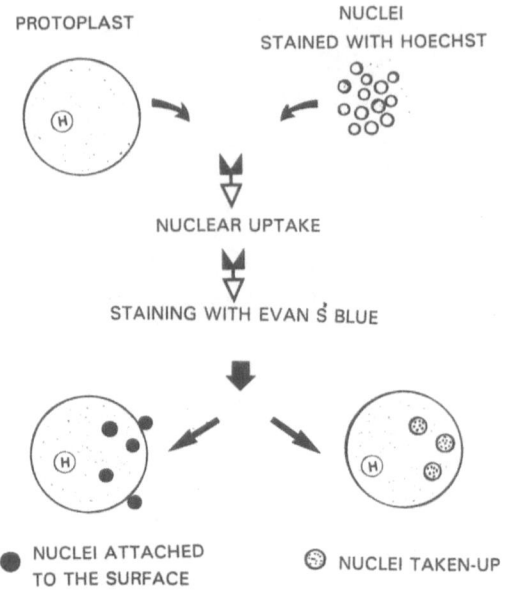

Fig. 2. A diagram showing the technique of differential staining to determine the frequency of nuclear uptake. H denotes the host nucleus.

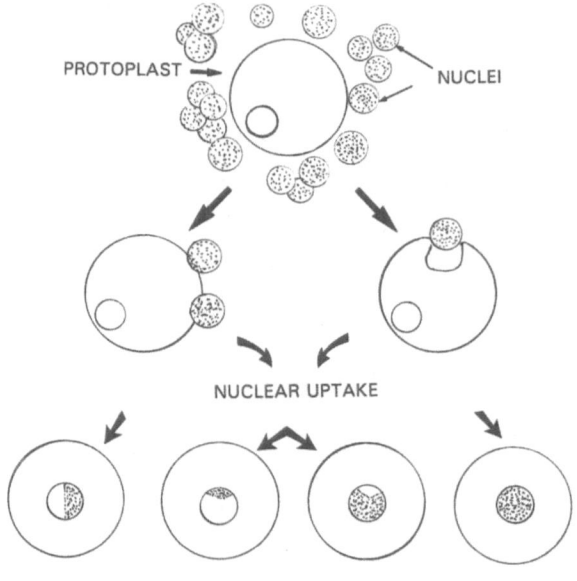

Fig. 3. A diagram showing the possible mechanisms of nuclear uptake and integration.

PTCM-D7/8

Procedures

The protocols described here are based on our studies with isolation of nuclei from protoplasts of *Vicia hajastana* and *Brassica nigra*, and their transplantation into the protoplasts of a pantothenate-requiring auxotroph of *Datura innoxia (Pn-1)* [22, 24].

Protocol 1. Isolation of protoplasts

Steps in the procedure

1. Arrange exponential phase (2 to 3-day-old) cell suspension cultures.
2. Using vacuum filtration, collect the cells over Miracloth and transfer 2–2.5 g of *Pn-1* cells to 10 ml of enzyme solution A dispensed into a 10 cm diameter Petri dish.
3. Use 10 ml enzyme solution B to digest 2–2.5 g *B. nigra* or *V. hajastana* cells. Prepare five Petri dishes.
4. Incubate the dishes for 2–3 h on a horizontal shaker (50 rpm) at 25 °C in the dark. Periodically examine the dishes under the microscope to ascertain complete cell wall digestion. When complete digestion had occurred, filter the suspensions through 85 μm nylon mesh and centrifuge at 120 × g for 5 min. Discard the supernatant. Suspend the pellet in 5 ml of 0.6 M sucrose, and layer 1 ml of mannitol solution on top of sucrose. Mannitol concentration for protoplasts to be used later for nuclei isolation should be 0.6 M and for recipient protoplasts it should be 0.5 M. Centrifuge for 5 min at 120 × g.
5. Protoplasts should gather at the interface of sucrose and mannitol. Remove the protoplasts with a Pasteur pipette and dilute the protoplast suspension with 8–10 ml of 0.5 or 0.6 M mannitol and centrifuge for 5 min at 120 × g.
6. Suspend the resulting pellet in appropriate mannitol solution (0.5 or 0.6 M).

Cell lines

— *Datura innoxia* P. Mill (*Pn-1*) cell suspension — a pantothenate-requiring auxotroph
— *Vicia hajastana* and *Brassica nigra* cell suspension cultures

Solutions
— Enzyme solution A: (in a 0.5 M mannitol solution)
    — 1% Cellulase Onozuka "R-10"
    — 1% Cellulase "RS"
    — 0.5% Macerozyme
    — 0.5% Rhozyme "HP150"
    — 5 mM Calcium chloride

- — Enzyme solution B: (in a 0.6 M mannitol solution)
  - — 1% Cellulase Onozuka ''R-10''
  - — 0.5% Cellulase ''RS''
  - — 0.5% Driselase
  - — 0.5% Macerozyme
  - — 0.5% Rhozyme ''HP-150''
  - — 5 mM Calcium chloride

Centrifuge the enzyme mixtures at 2500 × g for 15 min at 4 °C, collect the supernatant, adjust the pH to 5.7 and filter sterilize the solutions.
- — A solution of 0.6 M sucrose with 5 mM calcium chloride.
- — A solution of 0.6 M mannitol.
- — A solution of 0.5 M mannitol.

Protocol 2. Isolation of nuclei

Steps in the procedure

1. Centrifuge the tubes containing protoplasts for nuclei isolation for 5 min at 120 × g. Transfer 15 ml of ice-cold NIB to centrifuge tubes each containing 0.2 ml of pelleted protoplasts. Gently mix the suspension by tilting the tubes and leave for 5—7 min on ice.
2. Homogenize the suspension in a glass homogenizer by applying 10—15 gentle strokes and filter the homogenate through two layers of Miracloth, pre-soaked in NIB. This step will remove large debris and partially rupture protoplasts.
3. Filter the partially clean filtrate through a polycarbonate filter with a pore size of 12 μm (filter pore size should be chosen according to the size of nuclei which would vary from system to system).
4. Filter the suspension again, this time using a smaller mesh size (in this case 10 μm as the nuclei are about 8 μm in diameter, hence the decision to employ 12 and 10 μm mesh). Centrifuge at 120—150 × g. The selection of g force for centrifugation will depend upon the nuclear size. Transfer the supernatant to fresh tubes and centrifuge again. This two step centrifugation would ensure a higher yield of nuclei.
5. At this point, pool nuclear pellets from different tubes and add 5 ml of NIB-1. Centrifuge, discard the supernatant, and add 5 ml of NIB-1 again. Centrifuge to obtain the pellet composed of relatively pure assembly of nuclei.

Notes
5. If the final pellet is off-white, it will indicate that nuclei are more or less free of starch grains, but a white pellet indicates the presence of higher numbers of starch grains.
   If the nuclei are prepared for uptake experiments, the presence of starch grains can be ignored, but if desired, removal of starch grains can be achieved by 2 to 4 washes with NIB-1 employing centrifugation.
   Isolated nuclei can be stored in the refrigerator at 4 °C for later use or can be kept on ice for immediate experimentation.
   This procedure yields nuclei which are intact, biologically active, and free of cytoplasmic contamination (Fig. 1A)

Solutions
— Nuclei Isolation Buffer (NIB):
  — 10 mM MES
  — 0.2 M sucrose
  — 0.02% Triton X-100
  — 2.5 mM EDTA
  — 2.5 mM dithiothreitol
  — 0.1 mM spermine
  — 10 mM Sodium chloride
  — 10 mM Potassium chloride
Adjust the pH to 5.3, filter sterilize and store at 4 °C.

At this temperature this buffer can be stored for a week.
— NIB-1: Same buffer without the Triton X-100.

Protocol 3. Uptake of nuclei

Steps in the procedure
1. Obtain a density of about 2×10^6/ml of the recipient protoplasts (in this example, *Pn-1*) and a density of 10^8/ml of the nuclei (*B. nigra* or *V. hajastana*, the donor in this example), and mix the two suspensions to achieve a ratio of 1:50 (protoplast:nuclei) to have a better uptake, but the system can accommodate a ratio as low as 1:20 (protoplast:nuclei) without seriously jeopardizing the experiment.
2. Transfer a drop (500 µl) of protoplast/nuclei mixture to a 35 mm diameter Petri dish and wait for 5 min and then add 100 µl of UIS at the periphery of the mixture drop. Add four more drops (100 µl each) of UIS in a similar fashion at 30 sec. intervals, and incubate for 15–20 min.
3. Following the incubation, gradually dilute the UIS with 10 ml of 0.35 M mannitol, using 2 ml aliquots every 5 min.
4. Centrifuge the suspension at $120 \times g$ for 5 min.
5. Purify the protoplasts over 0.6 M sucrose, as described earlier, and suspend in 0.5 M mannitol.

Solutions
— Uptake Inducing Solution (UIS):
    — 30% Polyethylene glycol
    — 0.1 M Calcium chloride
    — 5% dimethylsulphoxide
Adjust the pH of this solution to 6.8.
— 0.35 M Mannitol with 5 mM calcium chloride.

Protocol 4. Determination of nuclear uptake frequency

Steps in the protocol
1. Stain isolated nuclei with 0.02% Hoechst''33258'' for 5–10 min at 0 °C and remove excess stain by washing 3–4 times with NIB-1.
2. Perform the uptake procedure as described above.
3. To a small volume (0.5–1 ml) of protoplast suspension from step 5 above, add an equal volume of 0.1% Evan's Blue solution. After 5 min, wash the suspension with 0.5 M mannitol to remove excess stain.
4. View a drop of suspension (placed on a glass slide with a cover-slip on) under the UV and BF of a fluorescence microscope. Use excitation filters UG I and BG 38 in combination with a barrier filter K 575.
5. The nuclei appearing blue and fluorescing will be the ones outside the protoplasts, whereas those taken up by protoplasts will show only the fluorescence and no blue colour. Score at least 500 protoplasts to calculate the frequency of nuclear uptake.

Solutions
- 0.02% Hoechst ''33258'' to be stored at 4 °C in the dark.
- 0.1% Evan's Blue in 0.5 M mannitol.

Protocol 5. Selection of prototrophic clones

Steps in the Protocol

1. Culture the protoplasts after nuclear uptake at a density of $10^5 \, ml^{-1}$ in 2.5 ml of MS-1 medium and after a week add 2 ml of MS-2 medium.
2. Transfer the cultures containing growing cell colonies to centrifuge tubes, allow to stand for 15 min, and remove the supernatant. Adjust the density of cell colonies to approx. 1000/ml of MS-2 medium.
3. Transfer 1–2 ml of the suspension obtained in step 2 to selection plates. To prepare selection plates, collect cells from a 2 to 3-day-old fast growing wild-type cell suspension culture (*Datura innoxia* in this example), suspend them in MS-2 at a density of 200 mg/ml and transfer 1 ml aliquots to 10 cm diameter Petri dishes containing 25 ml of MS-3. Swirl the Petri dishes gently to distribute the suspension evenly and after 5 h place a snugly-fitting filter paper (10 cm) on top of the cells. Place another filter paper (7 cm) on top of the first one and plate protoplast-derived colonies on this filter paper. Plating on filter paper facilitates the transfer of colonies from one plate to another. Black filter paper may be used to discern the surviving colonies easily. Isolate surviving prototrophic colonies after 2–3 weeks and subculture on MS-3 medium. These are putative transformed clones.

Solutions
- MS-1, MS [19] medium containing 0.2 M mannitol, 3% sucrose, 1 mg/l 2,4-D, 0.5 mg/l benzyladenine, and 0.5 mg/l Ca-D-pantothenate.
- MS-2, same as MS-1 but lacking Ca-D-pantothenate.
- MS-3, same as MS-2 but containing 0.8% Difco agar.

References

1. Anderson JW (1986) Extraction of enzymes and subcellular organelles from plant tissues. Phytochemistry 7: 1973–1988.
2. Becher D, Conrad B, Bottcher F (1982) Genetic transfer mediated by isolated nuclei in *Saccharomyces cerevisiae*. Current Genetics 6: 163–165.
3. Dunham VL, Bryant JA (1983) Nuclei. In: Hall JZ, Moore AL (eds) Isolation of membranes and organelles from plant cells, pp 237–275. New York: Academic Press.
4. Ferenczy L, Pesti M (1982) Transfer of isolated nuclei into protoplasts of *Saccharomyces cerevisiae*. Curr Microbiol 7: 157–160.
5. Fournier REK (1982) Microcell-mediated chromosone transfer. In: Shay JW (ed) Techniques in somatic cell genetics, pp 309–327. New York London: Plenum Press.
6. Fowke LC (1986) Ultrastructural cytology of cultured plant tissues, cells, and protoplasts. In: Vasil IK (ed) Cell culture and somatic cell genetics of plants, pp 323–342. New York: Academic Press.
7. Fowke LC, Gamborg OL (1980) Applications of protoplasts to the study of plant cells. Int Rev Cytol 68: 9–51.
8. Griesbach RJ (1987) Chromosome-mediated transformation via microinjection. Plant Sci 50: 69–77.
9. Hadlaczky G, Bisztray G, Praznovszky T, Dudits D (1983) Mass isolation of plant chromosomes and nuclei. Planta 157: 278–285.
10. Hughes BG, Hess WM, Smith MA (1977) Ultrastructure of nuclei isolated from plant protoplasts. Protoplasma 93: 267–274.
11. Kao FT (1983) Somatic cell genetics and gene mapping. Int Rev Cytol 85: 109–146.
12. Kobza J, Edwards GE (1984) Isolation of organelles: Chloroplasts. In: Vasil IK (ed) Cell culture and somatic cell genetics of plants, Vol 1, pp 471–482. New York: Academic Press.
13. Kondorosi E, Duda E (1980) Introduction of foreign genetic material into cultured mammalian cells by liposomes loaded with isolated nuclei. FEBS Lett 120: 37–40.
14. Lörz H (1985) Isolated cell organelles and subprotoplasts – their role in somatic cell genetics. In: Dodds JH (ed) Plant genetic engineering, pp 27–59. London: Cambridge University Press.
15. Lörz H, Potrykus I (1976) Uptake of nuclei into higher plant protoplasts. In: Dudits D, Farkas GL, Maliga P (eds) Uptake of nuclei into higher plant protoplasts, pp 239–244. Budapest: Akadémiai Kiadó.
16. Lörz H, Potrykus I (1978) Investigations on the transfer of isolated nuclei into plant protoplasts. Theor Appl Genet 53: 251–256.
17. Melchers G, Labib G (1974) Somatic hybridization of plants by fusion of protoplasts. I. Selection of light resistant hybrids of "haploid" light sensitive varieties of tobacco. Mol Gen Genet 135: 277–294.
18. Mii M, Saxena PK, Fowke LC, King J (1987) Isolation of chromosomes from cell suspension cultures of *Vicia hajastana* Grossh. Cytologia 52: 523–528.
19. Murashige T, Skoog F (1962) A revised medium for rapid growth and bio assays with tobacco tissue cultures. Physiologia Plantarum 15: 473–497.
20. Potrykus I, Hoffmann F (1973) Transplantation of nuclei into protoplasts of higher plants. Z Pflanzenphysiol 69: 287–289.
21. Saxena PK, King J (1989) Isolation of nuclei and their transplantation into plant protoplasts. In: Bajaj YPS (ed) Biotechnology in Agriculture and Forestry, Vol 9, pp 328–342. Berlin Heidelberg: Springer-Verlag.
22. Saxena PK, Fowke LC, King J (1985a) An efficient procedure for isolation of nuclei from plant protoplasts. Protoplasma 128: 184–189.
23. Saxena PK, Liu Y, Mii M, Fowke LC, King J (1985b) High nuclear yields from protoplasts of several plants. J. Plant Physiol 121: 193–197.
24. Saxena PK, Mii M, Crosby WR, Fowke LC, King J (1986) Transplantation of isolated nuclei into plant protoplasts – A novel technique for introducing foreign DNA into plant cells. Planta 168: 29–35.

25. Saxena PK, Liu Y, King J (1987) Nuclear transplantation into protoplasts: Optimal conditions for induction and determination of nuclear uptake. J Plant Physiol 128: 451–460.
26. Sree Ramulu K, Verhoeven HA, Dijkhuis P (1988) Mitotic dynamics of micronuclei induced by amiprophos-methyl and prospects from chromosome-mediated gene transfer in plants. Theor Appl Genet 75: 575–584.
27. Szabados L, Hadlaczky G, Dudits D (1981) Uptake of isolated plant chromosomes by plant protoplasts. Planta 151: 141–145.
28. Tallman G, Reeck GR (1980) Isolation of nuclei from plant protoplasts without the use of a detergent. Plant Sci Lett 18: 271–275.
29. Willmitzer L, Wagner KG (1981) The isolation of nuclei from tissue-cultured plant cells. Exp Cell Res 135: 69–77.
30. Xiao W, Saxena PK, King J, Rank GH (1987) A transient duplication of the acetolactate synthase gene in a cell culture of *Datura innoxia*. Theor Appl Genet 74: 417–422.

Index

3

Oncidium
  – *in vitro* culture C1/3
Orchard grass: see *Dactylis glomerata*
Orchidaceae (Orchids)
  – clonal propagation C1/1–7
Organ culture A4/1–14
Organogenesis
  – apple B8/5, 6
  – conifers C3/2, 3, 7–9
Oryza sativa
  – embryogenesis A5/11
  – protoplasts B1/3; B2/1–17; B3/1, 8
  – transgenic: B1/3; B2/1–17; B3/1

Packed cell volume A3/16, 17
Palms (Arecaceae)
  – clonal propagation C2/1–14
Pantothenate: see 'vitamins'
Papaver sp.
  – chromosome isolation: D7/6
Paphiopedilum
  – *in vitro* culture C1/2
Paraffin wax
  – use during tissue sterilization A2/11–13
Paromomycin
  – in selection of transformed cells B4/4–6;
    B8/8
Pelargonium sp
  – virus detection: C6/3
Petunia sp.
  – genes isolated D2/2
  – pollen nuclease D2/1
  – transformation by Agrobacteria B4/1;
    B6/1; B8/5
Phalaenopsis sp.
  – *in vitro* culture C1/2, 3
Phleomycin
  – selective agent B8/8
Phloroglucinol
  – component of media B8/3
Phoenix dactylifera (date palm)
  – clonal propagation: see palms
Phosphinothricin (PPT)
  – selective agent: B4/2, 5, 6
Phosphinothricin acetyltransferase (PAT) (enzyme, gene) A7/2; B9/2
Physcomitrella patens
  – protoplasts: A10/2, 17
Picea spp. (spruces)
  – *in vitro* culture C3/1, 11, 12, 13
Picloram
  – component of media A1/4; B1/5; C3/11
Pinus spp. (pines)
  – clonal propagation C3/1, 3, 5–9

Pollen
  – standard for cell sorting D5/7, 9–11
  – transformation D2/1–14
Polyethylene glycol (PEG)
  – for direct gene transfer A1/13; A7/1, 2,
    7–9; B2/3; B3/11, 13, 14
  – for protoplast fusion A1/13; B9/1, 2;
    D3/2, 9; D4/1, 2, 9, 11; D5/13
  – for uptake of nuclei: D7/4, 13
Polyvinylalcohol (PVA)
  – in direct gene transfer A7/1
Polyvinylpyrrolidone
  – inhibitor of phenol oxidases: D7/3
Potato: see '*Solanum tuberosum*'
Protein A-linked immuno-electron microscopy
  (PALIEM)
  – in virus testing: C6/4, 9
Protein determination B4/19
Protoplasts
  – chromosome preparations C4/1, 11
  – fusion (chemical) D3/1, 9; D4/1–17;
    D5/1, 13, 14; D7/2, 4–8, 13
  – fusion (electrical) A10/2, 3, 6, 8, 17;
    D3/1–11; D4/1
  – micromanipulation: A10/1–28
  – regeneration A2/2; A7/1–20; A9/2;
    B1/1–16; B2/1–17; B3/1–15; B9/1–13
  – source material A1/14–17; A4/2; A5/11;
    B4/1–24; B7/1
  – stable transformation A7/1–20; B1/3;
    B2/1–17; B3/1–15; B4/3; B7/1
  – wall-digesting enzymes A1/15; A3/20;
    A7/3, 7; B1/13; B2/14; B3/3, 4; B4/16;
    B9/9; C4/9; D3/1; D4/5, 6; D5/3, 5, 6;
    D7/9, 10
  – see also: 'cybrids', 'nuclei, isolation and
    uptake', 'transient gene expression',
    'somatic hybrids'
Pseudotsuga (Douglas fir) C3/1, 5
Pyridoxine-HCl: see 'Vitamins'

Raphanus sativus
  – protoplasts: A10/19
Restriction fragment length polymorphism(s)
  (RFLP)
  – of organellar genomes D4/15, 16; D6/1–8
  – of somaclones: C5/1–18
RFLP: see 'restriction fragment length polymorphism'
Rhodamine-6G
  – antimetabolite D4/7
Rice: see *Oryza sativa*
Root culture A4/1–14
Rye: see *Secale cereale*